The Moment of Proof

The Moment of Proof

Mathematical Epiphanies

Donald C. Benson

NEW YORK   OXFORD
OXFORD UNIVERSITY PRESS
1999

Oxford University Press

Oxford   New York
Athens   Auckland   Bangkok   Bogotá   Buenos Aires   Calcutta
Cape Town   Chennai   Dar es Salaam   Delhi   Florence   Hong Kong   Istanbul
Karachi   Kuala Lumpur   Madrid   Melbourne   Mexico City   Mumbai
Nairobi   Paris   São Paulo   Singapore   Taipei   Tokyo   Toronto   Warsaw

and associated companies in
Berlin   Ibadan

Published by Oxford University Press, Inc.
198 Madison Avenue, New York, New York 10016

Oxford is a registered trademark of Oxford University Press

Figure 15.9 on page 226 courtesy of B.B. Mandelbrot, *The Fractal Geometry of Nature* (W.H. Freeman, 1982).

Library of Congress Cataloging-in-Publication Data
Benson, Donald
The moment of proof: mathematical epiphanies / by Donald Benson
p.   cm.   Includes bibliographical references and index.
ISBN 0-19-511721-2
1. Proof theory—Popular works. I. Title
QA9.54.B46   1999
511.3—dc21   97-52139

9 8 7 6 5 4 3 2 1
Printed in the United States of America
on acid-free paper

Contents

Acknowledgments

I would like to acknowledge the loving encouragement and assistance of my dear wife Dorothy. Without her help, this book would never have been written. I wish to thank her especially for reading and rereading the manuscript and making suggestions and criticisms from the point of view of a mathematical novice, as well as for her many invaluable stylistic insights.

I would like to thank Ned Black for reading the entire manuscript and offering many valuable suggestions.

I would also like to acknowledge the assistance and encouragement of Kirk Jensen at Oxford University Press and the helpful suggestions of Henry Alder, Eric Benson, and Sherman Stein.

This book, apart from the frontmatter, was typeset using LaTex.

The Moment of Proof

Introduction

Euclid alone has looked on Beauty bare.

—EDNA ST. VINCENT MILLAY, *Sonnet*

I hope in this book to communicate something of the experience and the joy of mathematical discovery. One need not be the *first* to make a discovery to experience this pleasure. Being first is only important if one is concerned with fame, but the joy of discovery can be just as intense even if one is not the first. For example, we can still feel the pleasure that Euclid (c. 300 B.C.) experienced when he discovered that there are infinitely many prime numbers.

In the popular literature of science, it is usual to tell *what* is true rather than *why* it is true. For example, a mathematician can tell us that there are infinitely many prime numbers. It is interesting to know that this is true, but it is much more exciting to learn *why*.

A preoccupation with mathematics could even be dangerous. Archimedes (287–212 B.C.) lost his life when a Roman soldier ran a sword through him because he was so absorbed with his mathematics that he failed to respond to a command. Warning: Some readers may become similarly engrossed.

Mathematical symbols may seem forbidding at first, but mathematics is not a private language reserved for members of a secret society, as it was for the Pythagoreans of the sixth century B.C. Unfortunately, some mathematicians have done little to correct this false stereotype—some may even have promoted it. One of the aims of this book is to remedy this misconception.

There are sudden insights in mathematics that we will call *mathematical epiphanies*. The response to a mathematical epiphany can range from a silent "Aha," to Archimedes' cry of "Eureka," as he ran naked through the streets of Syracuse on discovering the principle of buoyancy. All mathematicians have experienced with great pleasure innumerable mathematical epiphanies. In fact, without the pleasure of these experiences, most would never have become mathematicians. This book aims to share mathematical epiphanies with a wider readership that includes many who consider themselves outsiders to mathematics.

Mathematical epiphanies are found in mathematical proofs that exhibit a certain charm. Mathematicians have a word for it—*elegance*. An elegant proof must meet the following requirements.

- It must have *mathematical importance*. The solution of a chess problem may elicit an "Aha." In the world of chess it may be considered elegant; it may even be mathematics of a sort; but it lacks mathematical importance.

- It must be *subtle*, but not obscure.

- It must be *short*—one or two pages at most.

- It must have a *surprise*—if not in its conclusion, then in its execution, like a mystery story in which we know who the culprit is but don't know how the detective will catch him.

Reading mathematics is quite different from general reading. Both the student and the professional mathematician know that they must read a book on mathematics at a leisurely pace. The reader must have the patience to follow an extended logical argument. It is helpful to have pencil and paper at hand. Sometimes the best way to take possession of a difficult idea is to write it down.

This book is eclectic in its choice of subject matter; however, it provides a self-contained introduction to many topics of mathematics as well as an introduction to the intellectual pleasures of mathematics.

1. **Recalling.** These chapters are intended to assist the reader in recalling mathematical skills that may be currently unused. The timely and important topic of chaos is also discussed.

2. **Counting.** A medley of topics centered around the theme of enumeration.

3. **Balancing.** The lever has applications to geometry.

4. **Numbering.** The imaginary numbers are misnamed. The prime numbers are useful to spies.

5. **Winning.** The mathematics of strategy and planning for business and recreation.

Welcome to all, and please join me in this exploration of mathematical elegance.

Part I

RECALLING

Chapter 1

Reflections

In another moment Alice was through the glass, and had jumped lightly down into the Looking-glass room.
— LEWIS CARROLL, *Through the Looking Glass*

Before we offer the main course, we present four mathematical appetizers. In each of the following four examples, based mainly on the Law of Reflection, the solutions depend on surprising geometric insights that lie just below the surface.

The Color of the Bear

Ada tells Ben about a puzzle that she read in the newspaper, but Ben's solution jumps to a conclusion.

Ada: A hunter leaves his cabin and walks 1 mile south. Then he walks 1 mile east, and, finally, he walks 1 mile north to end up where he started, at his cabin. To his surprise, he finds a bear inside. What is the color of the bear?

Ben: The bear must be a white polar bear because the hunter returns to his initial point after traveling 1 mile south, 1 mile east, and 1 mile north. *This is possible only if the hunter starts his journey at the North Pole.*

Example 1.1. Ben's last assertion is not entirely correct because the path he describes is one of many possibilities. In addition to the North Pole, find all the other places where such a circuit could start. *Hint:* At the additional locations, it is more likely that the hunter finds a penguin in his cabin instead of a polar bear.

Solution. Figure 1.1(a) shows an example of such a circuit in the vicinity of the South Pole. The circle has a circumference of 1 mile with its center at the South Pole. The point Q is an arbitrary point on the circle. The point P is 1 mile north of the point Q. The circuit encircles the South Pole S.

There are other solutions. For example, the circle in Figure 1.1(a) can be replaced by a circle with a circumference of 0.5 mile. Then instead of encircling the

(a)   (b)

Figure 1.1. (a) The circle is centered at the South Pole, and its circumference is 1 mile. Start at point P. Travel south 1 mile to point Q. Travel east 1 mile around the circumference of the circle, encircling the South Pole S and returning to Q. Travel north 1 mile, returning to P. (Every direction away from the South Pole is north.) (b) The solid circles are two members of an infinite family of circles centered at the South Pole representing possible initial points of the circuit. These circles approach the dotted circle which has radius 1 mile.

South Pole S once, we encircle it twice before returning north to point P. Moreover, circuits are possible that encircle the South Pole any number of times. The totality of initial points for the circuit consists of the North Pole and an infinite family of circles centered at the South Pole as shown in Figure 1.1(b).

In Figure 1.1(a), the hunter follows the shortest path between P and Q. The next example, from the noted British puzzle creator, Henry Dudeney[1] (1857–1930), also deals with a shortest distance.

The Spider and the Fly

Example 1.2. As shown in Figure 1.2(a), in a room measuring 30 feet in length and 12 feet in both width and height, a spider is at the middle of the one of the end walls, 1 foot from the ceiling, and a fly is at the middle of the opposite wall, 1 foot from the floor. The spider observes that he can reach the fly, as shown in Figure 1.2(a), by crawling vertically 11 feet down the wall to the floor, straight across the floor 30 feet to the opposite wall, and vertically 1 foot up the wall to the fly—a total of 42 feet. Assist the spider by showing that there is a path that is 2 feet shorter.

Solution. Imagine that the room is made of cardboard. Figures 1.2(b)–(e) show four ways to cut the box along certain of its edges in order to flatten it. *On the flattened box, the shortest path must be a straight line connecting* S *and* F. Figures 1.2(b)–(e) show the lengths of four different line segments corresponding to the four ways of flattening the box. The path in Figure 1.2(b) is the one contemplated by the spider, but the paths in Figure 1.2(d) and (e) are shorter. In fact, the path in Figure 1.2(e) is exactly 40 feet.[2] Surprisingly, as shown in Figure 1.2(f), the shortest path meets five of the six sides of the room.

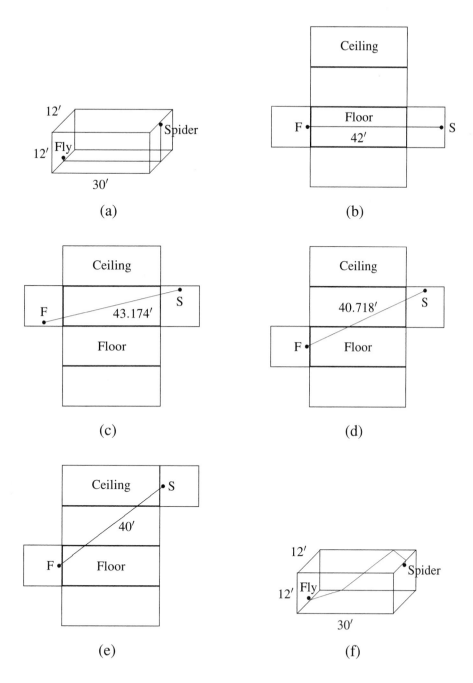

Figure 1.2. The Spider and the Fly. The spider wants to find the shortest path to the fly. The path shown in (a) and (b) is 42′, but it is not the shortest. Two other alternative paths are shown in (c) and (d). The shortest path (40′) is shown in (e) and (f).

The Law of Reflection

This search for the shortest path leads to a physical principle called the Law of Reflection. Figure 1.3 depicts the reflection of a light ray emanating from point A, reflected at point C by a mirrored surface l, and arriving at point B. The Law of Reflection states that *the angle of incidence θ is equal to the angle of reflection φ*.

The Law of Reflection takes a simpler form if we give substance to the "illusion" of reflection. In Figure 1.3, the portion of the figure above the line l shows the physical world in which photons bounce off a mirror. The portion below the line l is a representation of the illusion of reflection—Alice's "Looking-glass House"—in which the half plane above the line l is flipped across that line. After it strikes the mirror, it is useful to represent the further progress of the ray AC by the segment CB′, the mirror image of CB. *The fact that the angles θ and φ are equal implies that* ACB′ *is a straight line.* Since ACB′ is the shortest path connecting A and B′, it follows that ACB is the shortest path between A and B that meets the line l. *The reflected ray always follows the path of minimum length among all paths that meet l and connect the fixed points A and B.* This is one of the simplest of the far-reaching minimum principles that play a fundamental role in almost every aspect of physics.

The Law of Reflection also governs the reflection of sound and the motion of particles under elastic impact, as in the game of billiards.

Billiards

The traditional billiard table has dimensions 73.5 inches by 144 inches. We simplify the game by assuming that the cue ball is a moving point that moves in straight lines and that the ball bounces off the sides of the table, called *cushions*, according to the Law of Reflection.

Example 1.3. Suppose that the cue ball is placed at a point 46 inches from the west

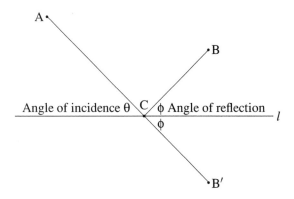

Figure 1.3. The Law of Reflection. The angle of incidence is equal to the angle of reflection.

cushion and 20 inches from the south cushion, and suppose that the ball returns to its initial position after striking, in order, the north cushion, the east cushion, and the south cushion. How far does the ball travel to return to its initial position?[3]

Solution. In billiards, a rebound from a cushion is similar to optical reflection except that there is no illusion of a reflected image. Nevertheless, in Figure 1.4, it is useful to show the billiard table and its repeated reflections across the edges of the table. The ball starts at point P, the solid dot; the five hollow dots represent the reflected images of the initial position of the ball. When the ball strikes the north cushion at Q, it bounces along the path QR. Figure 1.4 shows the image QR' of QR in the image of the billiard table reflected across its north edge. The reflected image is useful because, according to the Law of Reflection, PQR' is a straight line. Similarly, R'S' is the double reflection of RS; and S'P' is the triple reflection of SP. The reflected image PP' of the entire path of the ball is a straight line. Thus, the length of the circuit of the ball is equal to the length of the line segment PP', which is equal to 245 inches.[4]

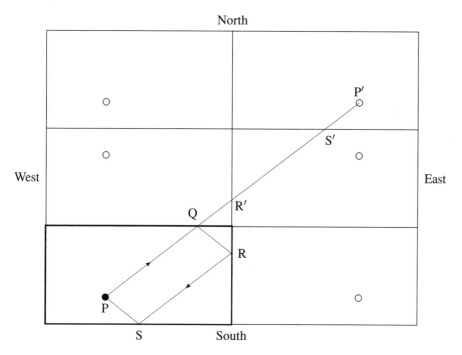

Figure 1.4. The bottom left box is the billiard table; the point P, the solid dot, is the initial position of the cue ball; and the parallelogram is the path of the ball. The remaining five boxes are obtained by reflecting the table repeatedly across its edges. The hollow dots represent the reflected images of the initial position of the cue ball. The length of the line segment PP' is the distance traveled by the cue ball.

The fourth and final example is an application of the Law of Reflection to a problem of geophysical exploration.

An Ear to the Ground

A *seismic reflection survey* is a method for mapping structures deep within the earth. Multiple *energy sources* and *listening devices* are employed. The energy source can be an earthquake, an explosion, or a device called a *vibroseis truck*, a vehicle weighing more than 10 tons that raises itself off the ground on a hydraulic plunger and vibrates the ground in the subsonic range of 10 to 56 cycles per second. Typically, four vibroseis trucks are used. The listening devices are called *geophones* and are used in an array of hundreds. This technique is used to map structures as deep as 30 miles below the surface of the earth.

The following example utilizes much simpler equipment—one seismic pulse and one listening device.

Example 1.4. Assume that there is a plane reflective layer of rock at an unknown depth below the plane surface of the earth and that the earth above the rock layer is homogeneous so that sound travels only in straight lines. Our trained elephant stamps his foot once and hears exactly two echoes from within the earth. What can be said about the angle between the plane of the rock layer and the plane of the earth's surface?[5]

Solution. Sound is propagated in rays in every possible direction. These rays are reflected by the rock layer and possibly re-reflected by the earth's surface according to the Law of Reflection. When a reflected ray returns to the earth's surface at the elephant's location, then he hears an echo.

Multiple echoes are caused by multiple reflections. If the rock layer is parallel to the earth's surface, then multiple echoes resemble the multiple images from parallel mirrors as is sometimes seen in a barber shop. I recall as a child sitting in a barber chair for the first time and watching an infinity of images of my small self. If the rock layer is parallel to the earth's surface, then an indefinite number of reflections is possible; the number of echoes heard is limited only by the elephant's ability to hear faint sounds. However, if the rock layer and the earth's surface are not parallel, then the number of echoes is limited regardless of the keenness of the elephant's hearing. For example, if the rock layer is a vertical wall, then no echo whatever is possible because no reflected ray can return to the earth's surface.

Figure 1.5(a) shows how it is possible for the elephant to hear two echoes from two reflected rays in case the rock layer is at an angle of 30° with the earth's surface. Point E represents the elephant's location. The first echo is reflected from the rock layer at A directly back to E, but the second echo is reflected at B by the the rock layer, then at C by the surface of the earth, and, finally, again at B by the rock layer. In both cases, the ray retraces its path to return to E.

In Figure 1.5(b), the lower half-plane is dissected into six wedge-shaped regions. The first region is bounded by the rock layer, tilted at 30°, and the surface of the earth. As we proceed counterclockwise, each region is the reflection of its predecessor across their common boundary. The points E_1, E_2, and E_3 are the reflected images of the elephant's location E.

First echo. In Figure 1.5(b), the line segment EA is the downward path of the first

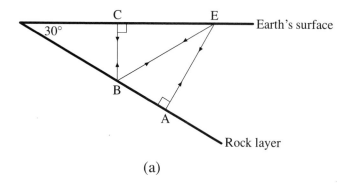

(a)

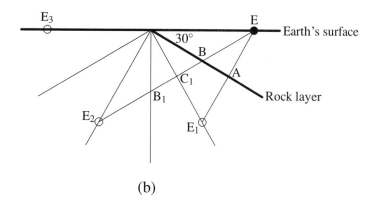

(b)

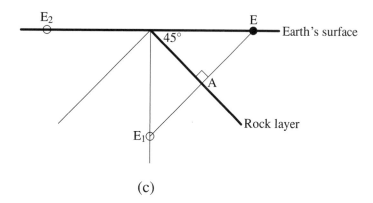

(c)

Figure 1.5. Seismic reflections.

echo, and line segment AE_1, the reflected image of AE in Figure 1.5(a), represents the reflection from the rock layer back to the elephant. It follows from the Law of Reflection that the segments EA and AE_1 fit together to form the straight line segment EE_1.

Second echo. The line EB is the downward path of the second echo; BC_1 is the reflection of the *upward* path BC in Figure 1.5(a); C_1B_1 is the image of the *downward* path CB; and, finally, B_1E_2 is the image of the *upward* path BE in Figure 1.5(b). Again, it follows from the Law of Reflection that, in Figure 1.5(b), the four segments EB, BC_1, C_1B_1, and B_1E_2 fit together to form the straight line segment EE_2.

The line segments EE_1 and EE_2 represent two echoes, but the line segment EE_3 does not represent an echo. In fact, a line from E to a reflected image of E represents an echo *only if that line proceeds downward from* E. For example, the line segment EE_2 does not represent an echo because it does not proceed downward from E. However, if the angle between the rock layer and the earth's surface were less than $30°$, then E_3 would be below the surface of the earth and EE_3 would represent a third echo. We conclude, if at most two echoes are heard, then the angle must be $30°$ or greater.

Figure 1.5(c) shows a rock layer that meets the surface of the earth in an angle of $45°$. The figure shows multiple reflections of the wedge-shaped region between the rock layer and the earth's surface; the first reflection is across the rock layer, the second reflection is across the vertical image of the earth's surface, and so on. The point E represents the position of the elephant—the point at which the echoes are heard. The points E_1 and E_2 are the images of E in the reflected regions. The straight line segment EE_1 represents an echo that is reflected from the rock layer directly to the earth's surface, but EE_2 does not represent an echo because it does not proceed downward from E. However, if the angle were less than $45°$, then the point E_2 would be below the surface of the earth and the line EE_2 would represent a second echo. We conclude, if two or more echoes are heard, then the angle is less than $45°$.

From the above two paragraphs, it follows that if the elephant hears exactly two echoes, the angle must be at least $30°$ but less than $45°$.

In this chapter, we have seen some mathematical insights without the use of mathematical formulas. Nevertheless, formulas, though not an end in themselves, are a useful mathematical supplement to ordinary language. There are many beautiful things that we can see more readily with just a bit of algebra. For example, in the next chapter, an equation helps to solve a problem that would otherwise be difficult.

Chapter 2

Hand in Hand

As the sun eclipses the stars by his brilliancy, so the man of knowledge will eclipse the fame of others in assemblies of the people if he proposes algebraic problems, and still more if he solves them.
—BRAHMAGUPTA (Indian mathematician, A.D. 598–670)

First, we examine the archery paradox of the ancient Greek philosopher Zeno of Elea (490 – 430 B.C.). Zeno's outrageous conclusion, that motion of any kind is impossible, contradicts the evidence of our senses. Here we show that Zeno's paradox also fails logically under mathematical thinking that today is elementary but was out of reach in Zeno's time.

Then, Zeno's paradox leads to a puzzle that asks, At what time are the hands of a clock precisely lined up? Algebra enables us to discover the answer.

Zeno's Paradox

Zeno's paradox states that an arrow can never reach its target. Zeno's idea is that the distance between the archer and the target can be dissected into infinitely many smaller and smaller distances. Before arriving at the target, the arrow must traverse each of these subintervals. Therefore, the total time of flight must be the sum of the times to traverse each of these subintervals. Because there are infinitely many of them, Zeno claims that the total flight time of the arrow must be infinite—in other words, the arrow never reaches its target.

The manner of dissection of the distance between the archer and the target into infinitely many subintervals is immaterial to Zeno's argument, but let us look at the consequences of one particular dissection. For simplicity, let us suppose that the distance from the archer to the target is one unit—since inches are too small and miles too large, let us say 1 rod (16.5 feet) from the target, and let us say that the arrow moves at a constant speed of 1 rod per second. Zeno claims that an arrow can never reach its target because to do so it must first travel nine-tenths of the distance to the target, then nine-tenths of the one-tenth rod remaining, then nine-tenths of the one-hundredth rod remaining, and so on forever. Since a finite time is required to traverse each of these distances, and since there are infinitely many of them, it follows, according to Zeno, that the arrow can never reach the target.

A sum with infinitely many terms is called an *infinite series*. A sum of finitely many terms of an infinite series is called a *partial sum*. The flaw in Zeno's argument is that he believes that every infinite series with positive terms must have an infinite sum. He does not realize that, given an infinite series with positive terms, one of *two* possibilities must be true:

1. There is a finite number that is larger than all of the partial sums.

2. There is no such number. In other words, there are partial sums that are as large as one pleases.

In the second case, we say that the infinite series *diverges to plus infinity* $(+\infty)$. Zeno mistakenly believes that this is the only possibility.

In the first case, we say that the infinite series *converges*. In this case, as we add more and more positive terms we get closer and closer to a certain finite number that is called the sum of the infinite series. In Zeno's archery example, it is clear that this finite number is, in fact, equal to 1 because the arrow reaches its target in 1 second. Therefore, 1 is the sum of the infinite series

$$\frac{9}{10} + \frac{9}{100} + \frac{9}{1000} + \ldots \tag{2.1}$$

> Reference numbers in the text that are in parentheses, such as (2.1), refer to *displayed formulas*; the reference number is located at the right margin. Other items such as theorems or figures have reference numbers that are not in parentheses.

Zeno's argument seems less cogent to us today because we have a lifetime of experience with a system of representing numbers, the Hindu-Arabic *positional numeral system* with the base 10, which is far superior to the system used by even the most learned of the ancient Greek mathematicians. Zeno used Greek letters, α, β, γ for 1, 2, 3 and ι, κ, λ for 10, 20, 30, a notation that made arithmetic exceedingly awkward. Today, decimal fractions are learned in grade school, but they were unknown to Zeno. In particular, we are taught that formula (2.1) can be written as .9999..., and we agree, without realizing that we are glibly dealing with an infinite series, that .9999... is an infinite decimal for the number 1. Although I am sure that Zeno was sincere and not a trickster, our current school training makes Zeno's argument seem to be a sophistry.

We also learn that a quantity can be expressed using an *indefinite* number of decimal places—more decimals mean greater precision. This background makes us willing to accept infinite decimals. We are comfortable with the idea that there is a number called π that has *infinitely* many digits in its decimal representation: $3.14159\ldots$. The idea is correct, but our acceptance of it is glib because it is based on training, not on firm logic. In fact, mathematicians accepted these concepts uncritically until the last half of the nineteenth century when it was discovered that many of their ideas about numbers were not founded on firm logic.[1] There continue to be mathematicians who concern themselves with the foundations of mathematics—a field with a surprising amount of controversy.

In Chapter 15 we will return to the question of what numbers really mean, but meanwhile it is best to concentrate on how numbers behave and what we can do with them. We can do a great deal of mathematics without knowing exactly what numbers are. Indeed, the mathematical pioneers, Newton, Euler, and others, did not concern themselves with these questions.

Solving an Equation

Let us look at another version of Zeno's paradox. The clock on my wall says 10:00 A.M. At some time between now and 11:00 A.M., the hour hand and the minute hand will be precisely lined up. But how can that happen? When the minute hand catches up to where the hour hand is now, the hour hand has moved ahead somewhat, and so on.

The same flaw in reasoning remains. It is not true that the infinite sum of the lengths of these smaller and smaller time intervals is infinite. The archery example is more clear than the clock example because in the former case we know that the total should be 1. Now we undertake the problem to determine what the total should be in the clock example. Let us see if we can find out when the two hands of the clock point in exactly the same direction. This is a fine example to illustrate the power of algebra. With just a little algebra we can easily solve the clock problem, and without algebra we probably cannot solve it at all.

I distinctly remember the pleasure I experienced when, in my early teens, I first saw the solution to this problem. In fact, I think that this was my very first mathematical epiphany.

The mathematical model

Why not solve the problem experimentally by observing the hands on an actual clock? One objection is that the answer depends on the accuracy of our measurements, but that is not the primary objection. The actual clock is not the subject of our investigation, but rather a mathematical model of a clock—a Platonic ideal clock.

Similar mathematical models play an important role in science. Scientific progress is furthered by a cyclic process that alternates experimentation with the construction of mathematical models that explain or embody the results of the experiments. Further experiments test the model and lead to modifications of it. Mathematicians are primarily interested in the inner workings of the model rather than its applications to science and technology. Most mathematicians are pleased that mathematics has such applications, although British mathematician G. H. Hardy (1877–1947) was a notable exception.[2] However, most mathematicians agree at least partially with Hardy by practicing and enjoying mathematics *primarily* for its own sake.

Mathematical models also occur in geometry, where we make a distinction between the *true* lines and circles of our theorems and the inaccurate figures that we draw on paper—a fact that justifies the saying that geometry is the science of drawing correct conclusions from incorrect diagrams. The study of the ideal clock may lead to conclusions concerning actual clocks, but our reason for discussing the ideal clock here is to show what algebra can do.

What is an ideal clock? We are not interested in what an ideal clock *is*. We are only interested in what it *does*—and only in the context of our particular problem. We do not even require that this clock tell the time. We only need to describe what the hands do between 10 and 11 o'clock. The hands point to two positions on the dial that can be characterized as a number of minutes between 0 and 60—the minute hand and the hour hand both point to numbers. We will use abbreviations: m stands for *the-number-that-the-minute-hand-points-to* and h stands for *the-number-that-the-hour-hand-points-to*. Abbreviations like this that stand for quantities are called *variables*. Because the minute and hour hands are connected by the clock mechanism, these two numbers have a relationship. Whenever the minute hand travels the 60 minutes from 10:00 to 11:00, the hour hand travels from the number 10 (or 50 minutes) on the clock face to the number 11 (55 minutes)—a span of just 5 minutes; therefore, the minute hand travels twelve times as fast as the hour hand. Whenever the minute hand travels 12 minutes, the hour hand travels 1 minute; whenever the minute hand travels m minutes, the hour hand moves $m/12$ minutes.

The equation

But the hour hand has a big head start. The minute hand is initially at 12 on the clock face, that is, at 0 minutes. On the other hand, the hour hand is initially at 10 o'clock, in other words, at 50 minutes. We can write the relationship between the hour hand and the minute hand as follows:

$$h = 50 + \frac{m}{12} \tag{2.2}$$

Formula (2.2) reads, "The position (measured in minutes) of the hour hand is equal to 50 plus the position (in minutes) of the minute hand divided by 12." A formula like (2.2) that says that two quantities are equal is called an *equation*. There are three different kinds of equations, all of which are usually expressed using the equal sign $=$.

1. *An equation that must be either true or false* because it contains no variables—only constants; for example, $1+1=2$ (true) and $2+2=5$ (false).

2. *An equation that may be true for some values of the variables and false for other values*; for example, equation (2.2). Finding the values of the variables for which the equation is true is called *solving* the equation. We do not reject equation (2.2) as false, and therefore useless, merely because some values of the variables make the equation false. For example, in equation (2.2), putting $m = 12$ and $h = 1$ gives $1 = 50 + \frac{12}{12}$ which is clearly false. Although equation (2.2) is not true for every choice of values of the variables m and h, a clock has a mechanical device that ensures that the numbers pointed to by the minute hand and the hour hand, respectively, are always numbers m and h for which equation (2.2) is true. For example, the pair of values

$$m = 12 \text{ and } h = 51$$

describes a possible position of the clock hands.

3. *An equation that is true for all values of its variables*; for example, $2x = x+x$. Such an equation is called an *identity*. When we want to emphasize that an equation is an identity, we use the symbol $\equiv$, as in $2x \equiv x+x$. As remarked above, putting $m = 12$ and $h = 1$ in equation (2.2) produces a false equation that shows that equation (2.2) is *not* an identity.

> The formula
> $$2x \equiv x+x$$
> is read,
> $2x$ *is identical to x plus x.*

The following device may clarify what an identity is. A dozen is 12, but it is not so commonplace how many constitutes a *gross*. To be convinced that an equation involving x is an identity, one might show that it is true for $x =$ one gross *without making use of the fact that a gross is* 144. For example, to be convinced that $2x \equiv x+x$ is an identity, observe that two gross is just as many as one gross plus one gross.

The equal sign $=$ is also used for defining either a new variable or a value for a previously defined variable. These usages may be signaled with the word "put."

1. We can define a new variable x in terms of a previously defined variable y by saying, for example, "Put $x = y+1$." Sometimes a modification of the equal sign such as $\triangleq$ is used for this meaning. The variable defined is always on the left side of the equal sign. In math jargon, "*Put $x = y+1$*," means, "*We define x to be equal to $y+1$.*"

2. Define a value for a previously defined variable x by saying, for example, "Put $x = 5$."

The scope of a *put* definition is generally limited to the discussion in which it occurs. We should avoid memorizing more than is necessary. In particular, we don't generally memorize definitions that begin, "*Put ....*" However, a definition might be worth memorizing if it begins with one of the following phrases:

- *Define ...*

- *If ..., then we say ...*

- *By ..., we mean ...*

Solving the equation

As stated above, finding the values of the variables for which an equation (not an identity) is true is called *solving* the equation. The positions of the hands of the clock correspond to solutions of equation (2.2). Let us find an equation that is true when the hands point in exactly the same direction. Solving that equation provides a solution to our problem. Let us introduce a variable p for the position in minutes at which both hands point in exactly the same direction. Since p is both the position of the minute hand and the hour hand, we can replace both m and h by p in equation (2.2), obtaining the equation

$$p = 50 + \frac{p}{12} \qquad (2.3)$$

By solving equation (2.3) we also solve the clock problem.

Definition 2.1. Two equations involving the same variable(s) are said to be equivalent if the two equations are true for exactly the same values of their variables.

Our method of solution is to use methods of transforming an equation into an equivalent new equation. The following general principle for transforming equations enables us to solve equation (2.3).

Fact 2.2. *Transforming an equation by adding (or subtracting) the same quantity to (or from) both sides of the equation, or by multiplying or dividing both sides by the same* nonzero *quantity, produces a new equation that is equivalent to the old equation.*

We omit the proof of this plausible assertion. However, some remarks are in order concerning why we are allowed to multiply or divide both sides of the equation only by nonzero quantities. Multiplication of both sides of an equation by zero produces $0 = 0$, which is uninformative and clearly not equivalent to the original equation.

On the other hand, division by zero is undefined. To see why this is so we must discuss the meaning of division. Division is defined in terms of multiplication, the more fundamental operation. When we say $6 \div 2 = 3$, we mean that 3 is the number such that when we multiply it by 2 we obtain 6. In general, for any numbers a and b, $a \div b$ denotes the unique number c such that if we multiply c by b we obtain a. But if a is different from 0 and b is equal to 0, then we are in trouble because no matter

---

Theorem-like items, including **Definitions, Facts, Theorems, Propositions, Examples,** and **Problems** are designated by a boldface title at the left margin. We refer to these items by reference numbers (e.g., **Definition 2.1**) that *share in common a sequential order*. This is why, in this chapter, **Fact 2.2** follows **Definition 2.1** and **Fact 2.1** does not exist. To locate a reference to a theorem-like item, the reader merely searches for the *reference number* of the item, ignoring its specific type—Definition, Fact, and so on. This system is especially useful in the later chapters of this book in which there are a large number of items of different types. The reader will be spared the frustration that the author has experienced in reading mathematical articles in which theorem-like references of each type are numbered independently.

what number we choose for c, we always obtain $c \times b = 0$; this product can never equal a unless $a = 0$. On the other hand, if $a = 0$ then there is not a *unique* number c such that $c \times 0 = 0$ because *any* number c satisfies the equality $c \times 0 = 0$.

Division illustrates the concept of *inverse* operation. Division is said to be the inverse of multiplication. In a similar fashion, subtraction is the inverse of addition: $a - b$ is defined as the unique number c such that $c + b = a$. We do not need to prohibit subtraction by 0 because $a - 0 = a$ satisfies all of our requirements.

The following chain of equations illustrates the application of Fact 2.2.

$$2x + 1 = 5 \quad \text{is true} \tag{2.4}$$

$$\text{if and only if} \quad 2x = 4 \quad \text{is true} \tag{2.5}$$

$$\text{which in turn is true if and only if} \quad x = 2 \quad \text{is true.} \tag{2.6}$$

- We subtracted 1 from both sides of equation (2.4) to obtain equation (2.5).

- We divided both sides of equation (2.5) by 2 to obtain equation (2.6).

According to Fact 2.2, exactly the same values of the variables satisfy the first and the last equations in this chain of equations. We have formed a chain of equivalent equations starting with equation (2.4) and ending with an equation (2.6) that is more informative; in fact, (2.6) is the *solution* of equation (2.4). An important skill in algebra—a skill that requires considerable practice—is to use Fact 2.2 to form such a chain of equations starting with an equation to be solved and ending with a solution— also called a *root*—of that equation.

Since algebra is an extension of ordinary language, formulas are embedded in normal English sentences. An algebraic argument is a convincing logical argument in the ordinary sense. In mathematics—in algebra, in particular—we become convinced of the truth of an assertion through the same logical processes that we use in ordinary life.

Let us return to the clock problem. Our goal is to find an equation equivalent to (2.3) that looks like

$$p = \text{some number} \tag{2.7}$$

To solve an equation like (2.7) is to do nothing. To see that such an equation is its own solution, recall that solving an equation means finding all values of the variables for which the equation is true.

To solve equation (2.3) we first try to find an equivalent equation in which all the p's are on the left side. We do this by subtracting $p/12$ from both sides of equation (2.3) as follows:

$$p - \frac{p}{12} = 50 + \frac{p}{12} - \frac{p}{12} \tag{2.8}$$

Now simplify the left and right sides of this equation. The left side of equation (2.8) is simplified by making use of the following identity:

$$p \equiv \frac{12p}{12}$$

assertion. A statement. The word *assertion* is used more often in mathematics than in ordinary usage, but it does not have a special technical meaning. An assertion must have, at least implicitly, a subject and a verb.

satisfy. Mathematical usage of the word *satisfy* differs from ordinary usage. To say that a particular value of a variable (e.g., $c = 0$) satisfies an equation means that substituting the variable c with the number 0 *solves* the equation (i.e., it changes the equation into an identity).

unique. In mathematics *unique* means *one and only.* The common (but incorrect) usage with the meaning "extremely novel" does not apply in mathematics.

Now we can see that the left side of equation (2.8) is equal to

$$\frac{12p}{12} - \frac{p}{12}$$

that, in turn, is equal to

$$\frac{11p}{12}$$

The right side is clearly equal to 50. Using these transformations of the left and right sides of equation (2.8), we obtain

$$\frac{11p}{12} = 50 \qquad\qquad (2.9)$$

Note that we have achieved our intention to put all the p's on the left side of the equation.

The rest is easy. We multiply both sides of equation (2.9) by $12/11$ because

$$\frac{12}{11} \times \frac{11p}{12} \equiv p$$

and so finally we obtain

$$\frac{12}{11} \times \frac{11}{12}p \equiv p = \frac{12}{11} \times 50 = \frac{600}{11} = 54.545454\ldots$$

When the hands of the clock are together they point to 54 minutes plus a little more than half a minute.

We have used algebra—more precisely, elementary algebra—to solve the clock problem. In discussing this problem we found that we could forget about clocks and concern ourselves just with variables, formulas, and equations.

Although much mathematics was known in antiquity, algebra is less than 400 years old. Algebra was an intellectual tool unknown to Pythagoras, Euclid, Archimedes, and other mathematical geniuses of antiquity.

Zeno's paradox raised the question of the meaning of infinity. Later, in Chapter 10, we will see a rigorous mathematical theory of infinity. In the next chapter, we arrive at some mathematical insights and some remarks about the Louisiana Purchase by taking a different look at Zeno's paradox.

Chapter 3

Further Progression

... half-of-half could still be halved,
With limitless division less and less.
        —LUCRETIUS, *On the Nature of Things,* 50 B.C.

Population, when unchecked, increases in a geometrical ratio.
    —THOMAS MALTHUS, *An Essay on the Principle of Population,* 1798

In the previous chapter we discussed Zeno's paradox using facts about decimal fractions. However, if we change the paradox slightly, the method of Chapter 2 no longer applies. In fact, the more usual statement of the paradox is that an arrow can never reach its target because to do so it must travel *half* (instead of *nine-tenths*) the distance between the archer and the target, then half the remaining distance, then half of that, and so on. We could discuss this version of the paradox by using the binary (base 2) representation of numbers, but we prefer to discuss it using *geometric progressions*, a topic that will be used in later chapters.

Exponents

Let us write a formula that describes this altered version of Zeno's paradox. Remember that we assume the distance between the archer and the target is 1 rod and that the arrow travels at a constant speed of 1 rod per second. It follows that in any time interval, the distance traveled in rods is equal to the time elapsed in seconds. In the first of the specified increments, time elapsed and distance traveled are both equal to one-half, leaving a distance equal to $1/2$ rod remaining to the target; and in the second increment, time and distance are $1/2 \times 1/2$, leaving a distance equal to $1/2 \times 1/2$ to the target. During the nth increment, the time elapsed and the distance traveled are both equal to

natural number. The *natural numbers* are the numbers

$$1, 2, 3, \ldots$$

also known as the *positive integers* or the counting numbers.

integer. The *integers* are the whole numbers—positive, negative, and zero.

real number. The *real numbers* are the numbers, including the negative numbers, that can be represented as finite or infinite decimal fractions. (The real numbers include the integers.)

For a^n read "*a* to the *n*."
For d_n read "*d* sub *n*."

Notation. Subscripts, for example, the 3 in d_3, are used to name variables that are related in some way, such as

$$d_1, d_2, d_3, \ldots$$

Multiplication of *a* and *b*, the product of *a* and *b*, is written in three different ways:

ab or $a \times b$ or $a \cdot b$

a and *b* are called *factors* of the product.

The *quotient* of a number *a* by a nonzero number *b* is also called a *fraction*, or *ratio*, and is written in the following ways:

$a \div b$ or a/b or $a : b$

$$\left(\frac{1}{2}\right)^n = \frac{1}{2^n}$$

and an equal distance remains to the target.

The *n* in 2^n is called an *exponent*. Let us take a few minutes to define in general what an exponent is. The notation 2^3 is an abbreviation of $2 \times 2 \times 2$. More generally, let *a* be an arbitrary real number and let *n* be a natural number; by a^n we mean the product

$$\underbrace{a \times a \cdots \times a}$$

Factor *a* repeated *n* times

The natural number *n* is called the exponent.

Now put d_n equal to the distance traveled (which is also equal to the time elapsed) during the first *n* increments. We have

$$d_n = \frac{1}{2} + \frac{1}{2^2} + \frac{1}{2^3} + \cdots + \frac{1}{2^n} \tag{3.1}$$

We will prove that

$$d_n = 1 - \frac{1}{2^n} \tag{3.2}$$

Q. How does this help explain the archery paradox?

A. It is helpful because from equation (3.2) it is fairly easy to show that the distance (or time) d_n is less than but very close to 1 if n is large.

If n is very large, then the last term of equation (3.2), $1/2^n$, is very small. In particular, one can easily verify on a hand calculator that if $n = 20$, then

$$\frac{1}{2^n} = \frac{1}{2^{20}} = \frac{1}{1,048,576} = 0.00000095367\ldots$$

In fact, we can make $1/2^n$ as small we please by taking n large enough. We could prove this more rigorously, but most readers would find further proof unnecessary for this plausible assertion. Since the distance traveled and the time elapsed in n steps is $1 - 1/2^n$, and since $1/2^n$ can be made as small as we please by taking n large enough, we see that the distance traveled and the time elapsed tends to 1 as n tends to infinity.

Q. Isn't the paradox still troubling because in (3.2) d_n tends to 1, but never actually reaches 1?

A. But, since time doesn't stop, the time actually reaches 1 second. When the time passes 1 second, then each and every one of these infinitely many time intervals has actually elapsed. Zeno's archery paradox is based on the tacit assumption that the sum of infinitely many nonzero time intervals must be infinite. In Chapter 2, we learned why Zeno's argument is flawed. The foregoing argument concerning the speed of the arrow shows that 1 is the sum of the infinite series

$$\frac{1}{2} + \frac{1}{2^2} + \frac{1}{2^3} + \cdots \qquad (3.3)$$

We will shortly see another proof of this fact when we show, using the formula for the sum of a geometric progression, that d_n tends to 1 as n tends to infinity, but first we need to discuss exponents.

Since geometric progressions have to do with exponents, we begin by discussing certain significant results called the *Laws of Exponents*. They are important, and we will make much use of them; therefore, we state them in the form of a theorem. We call it *Laws of Exponents—first version* because there are more general Laws of Exponents that we will discuss later.

Theorem 3.1 (Laws of Exponents—first version). *Let a be an arbitrary real number and let n and m be natural numbers.*

- Addition of Exponents.

$$a^n \times a^m = a^{n+m} \qquad (3.4)$$

- Multiplication of Exponents.

$$(a^n)^m = a^{nm} \qquad (3.5)$$

> **Notation.**
> $\neq$ means *not equal.*
> $>$ means *greater than.*
> $<$ means *less than.*
> $\geq$ means *greater than or equal.*
> $\leq$ means *less than or equal.*

- Subtraction of Exponents. *If $a \neq 0$ (we require $a \neq 0$ because division by 0 is not defined) and $n > m$, then*

$$a^n \div a^m = a^{n-m} \qquad (3.6)$$

Proof. To show that Addition of Exponents is true, notice that the following product is equal to both the left and the right sides of equation (3.4).

Factor a repeated $n + m$ times.

$$\overbrace{(a \times a \times \cdots \times a)} \times \underbrace{(a \times a \times \cdots \times a)} \qquad (3.7)$$

Factor a repeated n times.   Factor a repeated m times.

This proof of Addition of Exponents is short, but the prize for brevity really goes to the Indian mathematician Bhāskara (1114–c. 1185) who gave a proof that consisted of a diagram and the one word, "Behold!"

To prove the Law of Multiplication of Exponents, notice that the following product is equal to both the left and the right sides of equation (3.5).

m products, each containing n factors

$$\overbrace{\underbrace{(a \times \cdots \times a)}_{n \text{ factors}} \times \underbrace{(a \times \cdots \times a)}_{n \text{ factors}} \times \cdots \times \underbrace{(a \times \cdots \times a)}_{n \text{ factors}}}$$

The idea of Subtraction of Exponents is that in the fraction a^n / a^m we can cancel m a's from both the numerator and the denominator. To *prove* Subtraction of Exponents, put

$$n' = n + m \tag{3.8}$$

so that $n = n' - m$. Using this substitution, equation (3.4), which we have just established, becomes

$$a^{n'-m} \times a^m = a^{n'}$$

Now, since Subtraction of Exponents assumes $a \neq 0$, we can divide both sides of this equation by a^m and reverse the order of the equality to obtain

$$\frac{a^{n'}}{a^m} = a^{n'-m} \tag{3.9}$$

which is identical with equation (3.6) apart from the replacement of n by n'. In fact, equation (3.9) is valid for all natural numbers m and all n' such that $n' > m$. This is true because in equation (3.8), n can be any natural number, which implies that n' can be any natural number greater than m. Therefore, equation (3.9) says exactly the same thing as equation (3.6). □

> For n' read "n prime."
>
> The symbol n' is the name of a variable. This particular name is chosen to emphasize a relationship with the variable n.

Q. Why do you call this a proof? What is a proof?

A. There is some disagreement even among mathematicians as to what constitutes a proof. I like the following definition. A proof is a convincing argument, convincing, that is, to an honest skeptic with a real desire to learn the truth. In other words, proof in mathematics is the same as proof in everyday life; however, the standard is higher. For example, in a courtroom the highest standard is "proof beyond reasonable doubt," but mathematics requires proof beyond any doubt whatever.

It is not possible to construct a machine that tells us whether a certain argument is a proof because a machine cannot tell us whether a person will be convinced. Mathematical logic endeavors—with great success—to construct abstract systems that model the manner in which mathematicians convince one another. However, these systems are only shadows of the real thing.

Q. Are you saying that a computer cannot give a mathematical proof?

A. A computer can give all or part of a proof. If we are convinced, then it is a proof. To be convinced by a computer proof, we must (1) be convinced that the computer has been programmed to provide at least part of a convincing argument, and (2) that the software and hardware are free of errors that may affect the outcome of the argument.

Q. I may be convinced by an argument, but couldn't I be merely deluded?

A. Even with careful scrutiny, there is no way to entirely avoid the risk of accepting an incorrect argument. What was considered a proof yesterday may not be accepted as such today. However, in mathematics many arguments have stood the test of time. Some have been considered convincing by mathematicians for hundreds—even thousands—of years. Nevertheless, some arguments have been accepted as mathematical proofs for years and have finally been shown to be incorrect. A notable instance is the "proof" of the Four-Color Conjecture by A. B. Kempe in 1879. His proof was considered correct for eleven years, but was finally shown to be flawed.

The Four-Color Conjecture asserts that any map can be colored with no more than four colors in such a way that no two contiguous countries have the same color. In 1976, almost 100 years after Kempe's erroneous proof, Kenneth Appel and Wolfgang Haken announced a proof of the Four-Color Conjecture that has subsequently been accepted by experts in the field. The proof of Appel and Haken relies, in part, on a computer analysis of 1,936 special cases.[1]

Mathematicians often present proofs in a formal way. These formalities are a little like the formalities of a courtroom trial. There are conventions as to what should be presented to the jury and how it should be presented. Don't be intimidated by these formalities; just remember that you are the jury. The mathematical assertions to be proved are called theorems, propositions, lemmas, and corollaries. Proofs are delimited so that we know where the proof is supposed to start and where it is supposed to end. The beginning is often marked with the word Proof, and the end is marked with the abbreviation Q. E. D. (old fashioned) or the symbol □ placed at the right margin like this: □

theorem. A mathematical assertion to be proved.

proposition. A theorem usually of secondary importance.

lemma. A theorem that assists in the proof of another theorem.

corollary. A theorem that is an immediate consequence of another theorem.

Q.E.D. Latin. *Quod erat demonstrandum.* That which was to be proved.

Extended Laws of Exponents

The first version of the Laws of Exponents requires that exponents are natural numbers. Indeed, we have only defined the expression a^n in case n is a natural number. It is possible to extend this definition if we forego the possibility that a is negative.

Fact 3.2. *Given arbitrary real numbers a and b with a positive, the exponential a^b can be defined such that the following conditions hold:*

1. Assuming, as above, that a is positive, a^b is also a positive real number.

2. If b happens to be a natural number, then a^b is consistent with our previous definition. For example, a^3 is equal to $a \cdot a \cdot a$.

3. The Laws of Exponents (Theorem 3.1) hold with the following modifications:

   (a) *Weaker assumptions on the numbers m and n.* Theorem 3.1 requires that m and n be natural numbers. We relax this requirement so that m and n are permitted to be arbitrary real numbers. (Note that m and n are now permitted to be negative or zero.)

   (b) *Stronger assumption on the number a.* In Theorem 3.1, a was an arbitrary real number. Now we add the restriction that a must be positive.

The following examples show some consequences of Fact 3.2.

Example 3.3. Show that Fact 3.2 implies $a^0 = 1$.

Solution. From the Law of Addition of Exponents, we have

$$a^0 a^0 = a^{0+0} = a^0$$

In other words, a^0 is a number x that satisfies $x \cdot x = x$. The only numbers that satisfy this condition are 1 and -1. But $a^0 = -1$ is not a possibility because condition 1 above requires that a^0 is positive. Therefore, we must have $a^0 = 1$.

Example 3.4. Show that for every natural number n, we have

$$a^{-n} = \frac{1}{a^n}$$

Solution. From Addition of Exponents (3.4), it follows that we have

$$a^n \cdot a^{-n} = a^{n+(-n)} = a^0 = 1$$

We obtain the claimed result by dividing the right and left sides of this equation by a^n.

Example 3.5. Show that for every natural number n, we have

$$a^{1/2} = \sqrt{a}$$

In other words, $a^{1/2}$ is the unique positive real number whose square is a.

Solution. From Addition of Exponents (3.4), we have

$$a^{1/2} \cdot a^{1/2} = a^{\frac{1}{2}+\frac{1}{2}} = a^1$$

But from the consistency requirement 2 above we must have $a^1 = a$. Hence, $a^{1/2}$ must be the unique positive square root of a.

Similarly, $a^{1/3}$ is the cube root of a, and so on.

Example 3.6. Let p and q be natural numbers. Show that $a^{p/q}$ is

1. equal to the qth root of a^p, and

2. equal to the qth root of a raised to the pth power.

Solution. From Multiplication of Exponents (3.5), we have

$$\left(a^{p/q}\right)^q = a^{\frac{p}{q} \cdot q} = a^p$$

It follows that $a^{p/q}$ is equal to the qth root of a^p. On the other hand, again from Multiplication of Exponents (3.5), we have

$$\left(a^{1/q}\right)^p = a^{\frac{1}{q} \cdot p} = a^{p/q}$$

Since $a^{1/q}$ is the qth root of a, we see that this root raised to the pth power is equal to $a^{p/q}$.

In the study of calculus, one frequently puts $a = e$, where e is the constant $e = 2.71828\ldots$. The constant e, together with π, is one of the most important constants in mathematics. Figure 3.1 is a graph of e^x.

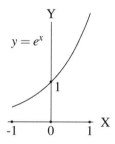

Figure 3.1. The exponential e^x.

Q. What is 0^0? Is it undefined, like $0/0$?

A. The choice whether to define 0^0 is based on convenience, not on correctness. If we refrain from defining 0^0, then certain assertions become unnecessarily awkward. For example, consider the following proposition:

If n is a positive integer and x is real number satisfying

$$0 \leq x \leq 1$$

then we have

$$x^{n-1} \geq x^n$$

If we consider 0^0 to be undefined, then we must add the additional assumption:

Either $n > 1$ or $x > 0$.

But if we define $0^0 = 1$, then we avoid this complication, and the proposition becomes true in case $n = 1$, $x = 0$. The consensus is to use the definition $0^0 = 1$, although there are textbooks that refrain from defining 0^0.

Geometric Progressions

We introduce geometric progressions through an examination of *compound interest,*
a concept that is doubtless familiar to the reader. A sum of money is invested at a
certain interest rate for a period of years. At the end of each year, interest is computed
not just on the initial sum, but also on the total interest accumulated to date.

Compound interest is a useful concept for making financial decisions. For example, it is generally agreed that President Jefferson got an outrageously good value
for the 15 million dollars paid to France in 1803 for the Louisiana Purchase, but how
do we go about making that judgment? It is too naïve to say that today this huge
part of the United States is worth vastly more than 15 million dollars because this
ignores what the French could do with the money over a period of almost 200 years.
Holding money at compound interest is the standard investment for making this kind
of comparison. If the French had invested the 15 million dollars, what would it be
worth today?

We must choose an interest rate. Since we suspect that the French got a bad deal,
perhaps we can afford to be generous here. Let's suppose that the French were able
to invest the money at 8%, compounded annually.

Let's write how much the investment was worth initially and at each anniversary
of the purchase. (We use $15M$ as an abbreviation for $15,000,000$.) At the end of the
first year it was worth

$$15M + 15M \times 0.08$$

*Q. This formula is a bit confusing. If I do the arithmetic from left
to right, the order in which I normally read text, then I must add $15M +$
$15M = 30M$ and then multiply this by 0.08, obtaining $\$2.4M$. But this
is less than the initial value of $\$15M$. What is wrong here?*

*A. Algebraic formulas are not evaluated from left to right. There is
a convention called the* order of precedence *that prescribes the order in
which the arithmetic operations are performed. In the absence of any
grouping symbols such as $(\ )$, $[\ ]$, or $\{\ \}$, arithmetic operations are done
in the following order.*

1. Exponentiation $(2^7, a^b)$

2. Multiplication

3. Division

4. Addition

5. Subtraction

The mnemonic for this is, "Excuse My Dear Aunt Sally."

Here is an example:

$$4 + 64/2 - 3 \times 2^3 = 4 + 64/2 - 3 \times 8 \quad \textit{First do the exponentiation,}$$
$$= 4 + 64/2 - 24 \qquad \textit{then the multiplication,}$$
$$= 4 + 32 - 24 \qquad \textit{then the division,}$$
$$= 36 - 24 \qquad \textit{then the addition,}$$
$$= 12 \qquad \textit{and finally the subtraction.}$$

The formula that gives the value of the investment at the end of the first year is evaluated as follows:

$$15M + 15M \times 0.08 = 15M + 1.2M \qquad \textit{First the multiplication,}$$
$$= 16.2M \qquad \textit{and then the addition.}$$

Q. Does the precedence convention compel me to do the multiplication first? What if I encounter another problem in which I really want to do the addition first?

A. In that case, you must use grouping symbols (parentheses, brackets, or braces) to override the normal order of precedence. Since parentheses override the order of precedence, there is an extended mnemonic: "Please Excuse My Dear Aunt Sally." (If there is more than one pair of grouping symbols, the inmost pair defines the group that is evaluated first.) The problem that you suggest, which is unrelated to our interest computation, needs parentheses as follows:

$$(15M + 15M) \times 0.08 = 30M \times 0.08 = 2.4M$$

Returning to the original problem, the value of the investment is multiplied by 1.08 at the end of the first year. In fact, at the end of *every* year the value of the investment is multiplied by this factor.

$$\text{value at end of year} = 1.08 \times \text{value at start of year}$$

The factor 1.08 is an example of an *annual multiplier*. In general,

$$\text{annual multiplier} = 1 + \text{interest rate}$$

To find the value of the investment at the end of the second year, we apply the annual multiplier twice:

$$\text{value at 2 years} = 1.08 \times \text{value at 1 year}$$
$$= 1.08 \times 1.08 \times \text{initial value}$$
$$= 1.08^2 \times 15M = \$17.496M$$

Continuing in this way, we see that the list of values of the investment at the yearly anniversaries is

1803	1804	1805		1999	
$15M,$	$15M \times 1.08,$	$15M \times 1.08^2,$	$\cdots,$	$15M \times 1.08^{1999-1803}$	(3.10)

The 1999 item represents the current value of the investment:

$$15M \times 1.08^{1999-1803} = 15M \times 1.08^{196} = \$53,351,586M$$

in other words, about 53 trillion dollars.

> Q. *Isn't it far-fetched to use this calculation to make a conclusion about the Louisiana Purchase?*
>
> A. *Although our calculation is rough, it supports the contention that the French were not cheated when they agreed to the Louisiana Purchase for only 15 million dollars. Indeed, since the current dollar value of the Louisiana Purchase is measured in trillions, not millions, our rough computation gives the right order of magnitude; this is remarkable considering that our computation assumes, without foundation, that there is no major discontinuity between the value of current dollars and 1803 dollars, and considering that the choice of 8% as the interest rate is arbitrary.*

Let us generalize this compound interest problem. Suppose that the initial investment $= a$, the annual multiplier $= r$, and the number of years $= n$. The list of anniversary values is an example of a *geometric progression*. More generally, a geometric progression is a list of numbers such that any two adjacent numbers have the same quotient, called the *common ratio*. Here is a formal definition.

Definition 3.7 (geometric progression). Let a (the *initial term*) and r (the *common ratio*) be arbitrary numbers and let n be a natural number. Then the sequence

$$a, ar, ar^2, ar^3, ar^4, \ldots, ar^n \tag{3.11}$$

is called a *geometric progression*.

For example, if $n = 2$, then the geometric progression is

$$a, ar, ar^2 \tag{3.12}$$

so that there are 3 terms. In general, the number of terms in the geometric progression is $n + 1$.

> Q. *What's geometric about this? Why is it called a geometric progression?*
>
> A. *The ancient Greeks were first to use this term. The most likely reason is that certain geometric magnitudes such as the length of the diagonal of a unit square, a one-by-one square, did not fit the Greek concept of number. Geometric magnitudes were considered conceptually different from arithmetic magnitudes. The term geometric progression was probably used because geometric magnitudes often arose in their discussion of sequences like (3.11). For example, the Greeks studied the problem of finding the middle term of (3.12) if the first and last terms are known. The middle term of this three-term geometric progression is called the geometric mean of the first and last terms. Explicit calculation of the geometric mean involves square roots, which the ancient Greeks considered geometric, not arithmetic, magnitudes.*

Note that equation (3.1) is the sum of a geometric progression. The formula for the sum of the geometric progression (3.11) is stated in the following theorem.

Theorem 3.8 (sum of a geometric progression). *For any numbers a and r and for any natural number n, the following equations are true:*

$$a + ar + ar^2 + ar^3 + \cdots + ar^n = \frac{ar^{n+1} - a}{r - 1} \qquad if\ r \neq 1 \qquad (3.13)$$
$$= (n+1)a \qquad if\ r = 1$$

Q. Where did the right side of equation (3.13) come from?

A. I ask that you defer asking this question until you look carefully at the proof below. You will see that the proof itself makes the right side of equation (3.13) natural and plausible. Mathematics is often written in this style. A theorem is stated, and then it is proved. The process is very orderly and facilitates the verification of logical correctness and the documentation of new research, but it can be very difficult for the novice because this style tends to conceal the discovery process. However, mathematicians know that there is much more to understanding a mathematical result than merely verifying its correctness. Reading mathematics is slow because we want not only to see that a result is correct, but also we wish to discover its underlying meaning. Mathematicians use two terms to describe the ways in which they present mathematics.

Rigorous *is used to describe a mathematical argument that emphasizes logical correctness.*

Heuristic *is used to describe an argument that emphasizes the underlying meaning, the discovery process, the motivation, and applications to mathematics, science, or ordinary life.*

It helps to understand Theorem 3.8 if we relate it to the Louisiana Purchase example. Consider the problem of finding the total amount of interest on the investment of $15M$.

We already found the total current value of the investment: $15M \times 1.08^{196}$. And we know that the original investment was $15M$. To obtain the total interest, we just subtract the original investment from the total value, obtaining

$$15M \times 1.08^{1999-1803} - 15M = 15M \times 1.08^{196} - 15M = 53,351,571M \qquad (3.14)$$

There is another way to look at this problem. The interest at the end of the first year is $15M \times 0.08$. At the end of the second year, the principal is $15M \times 1.08$; therefore, the interest the second year is $15M \times 1.08 \times 0.08$. This may seem tedious, but please bear with me. At the end of the third year the principal is $15M \times 1.08^2$ so that the interest at the end of the third year is $15M \times 1.08^2 \times 0.08$. Here is the sum that represents the entire interest payment:

1804	1805	$\cdots$	1999

$$(15M \times .08) + (15M \times 1.08 \times .08) + \cdots + (15M \times 1.08^{195} \times .08) \qquad (3.15)$$

Note that the exponent in the last term is 195 instead of 196 as it was in equation (3.10) because we are computing interest on the balance *before the addition of interest*. Thus, the total interest is the sum of a geometric progression. This becomes more clear if, using the fact that $.08 \times 15M = 1.2M$, we write the sum as follows:

$$\overset{\textbf{1804}}{1.2M} + \overset{\textbf{1805}}{1.2M \times 1.08} + \overset{\cdots}{\cdots} + \overset{\textbf{1999}}{1.2M \times 1.08^{195}} \qquad (3.16)$$

Equation (3.13) gives a way of evaluating this sum. In fact, put

$$n = 195, \quad a = .08 \times 15M = 1.2M, \text{ and } r = 1.08$$

The right side of equation (3.13) then becomes

$$\frac{1.2M \times 1.08^{196} - 1.2M}{1.08 - 1}$$

which is equal to $15M \times 1.08^{196} - 15M$.

Note that this is the very same result that we got in equation (3.14) using a simpler calculation. By showing that (3.16) is equal to (3.14), we have verified a special case of Theorem 3.8. Now let's prove the general case. Notice that the r on the right side of (3.13) is the *annual multiplier* and $r - 1$ is the *interest rate*.

Proof of Theorem 3.8. The case when $r = 1$ is obvious because in that case each term in the sum is equal to a, and there are $n + 1$ terms.

Dangerous curve! ↪   In case $r \neq 1$, put S equal to the left side of equation (3.13). The result of multiplying S by r is

$$rS = ar + ar^2 + ar^3 + ar^4 + \cdots + ar^{n+1}$$

Now we subtract S from rS. We write this subtraction in such a way that the terms are arranged in columns that cancel.

$$\begin{aligned} \text{From} \quad & rS = && ar + ar^2 + ar^3 + \cdots + ar^n + ar^{n+1} \\ \text{subtract} \quad & S = a + ar + ar^2 + ar^3 + \cdots + ar^n \end{aligned}$$

The result of this subtraction is

$$rS - S = ar^{n+1} - a \qquad (3.17)$$

But $rS - S$ is equal to $(r - 1)S$, and therefore equation (3.17) becomes

$$(r - 1)S = ar^{n+1} - a \qquad (3.18)$$

Finally, divide both sides of equation (3.18) by $r - 1$ to obtain equation (3.13).   □

Q. At the beginning of this chapter you promised to use geometric progressions to resolve a variant of Zeno's paradox. What does Theorem 3.8 have to do with the archery paradox?

A. Recall that in our discussion of the paradox earlier in this chapter, the proof of equation (3.2) was deferred. We can prove it now as a special case of Theorem 3.8.

Example 3.9. Use Theorem 3.8 to give a more succinct form to the right side of equation (3.1) by putting $a = 1/2$, and $r = 1/2$; and replacing n by $n-1$ in equation (3.13).

Solution. We obtain

$$\frac{1}{2} + \frac{1}{2^2} + \frac{1}{2^3} + \cdots + \frac{1}{2^n} = \frac{\frac{1}{2}\frac{1}{2^{(n-1)+1}} - \frac{1}{2}}{\frac{1}{2} - 1} = 1 - \frac{1}{2^n}$$

Thus we have shown that equation (3.2) is correct.

Here is another simpler application of Theorem 3.8.

Example 3.10. Let n be a natural number. Then put $a = 1$ and $r = 2$ to obtain

$$1 + 2 + 2^2 + 2^3 + \cdots + 2^n = 2^{n+1} - 1$$

The Logistic Equation and Chaos

According to Thomas Malthus, population tends to grow as a geometric progression. Malthus correctly pointed out the misery that results when population outstrips resources, but the geometric rate of population growth is moderating, at least in the industrialized countries. Currently, the population of the United States is growing at the rate of 0.6% per year, a geometric progression with the common ratio $r = 1.006$. If the population were truly a geometric progression, the rate of 0.6% would be constant, but, in fact, it is expected to be zero by the middle of the next century.

We can express geometric growth in the following manner. We number the years $0, 1, 2, \ldots$, and we denote the population in the nth year as x_n. If r is the common ratio, then geometric growth must satisfy

$$x_{n+1} = rx_n, \quad n = 0, 1, 2, \ldots \tag{3.19}$$

If the population x_0 in the first year is known, then the population in each succeeding year can be computed recursively from equation (3.19). Because the population in the first year determines the population at each succeeding year, geometric growth is called a *deterministic* mathematical model.

The model of geometric growth fails to account for a difficulty common to all populations, whether humans, coyotes, sparrows, wildflowers, or bacteria. When population grows beyond a certain size, growth stops or reverses due to competition for resources. The simplest mathematical model that reflects this situation is called the *logistic equation*. We choose units of population so that the population is always between 0 and 1. As in equation (3.19), the variables $x_0, x_1, x_2, \ldots$ represent the population at the beginning of each year. The logistic equation is the following:

$$x_{n+1} = rx_n(1 - x_n), \; n = 0, 1, 2, \ldots \tag{3.20}$$

Like the geometric model, the logistic model of population growth is also deterministic because the population x_0 in the first year recursively determines the population in each succeeding year. The positive constant r, called the *fecundity* of the population, depends on the particular case: the species, resources, and other conditions. If the factor $1 - x_n$ were replaced by a constant, then the population would grow (or decline) exactly in a geometric progression. If for some n, x_n is close to 1, then a condition of overpopulation causes a marked decline in the next time period. If x_n is close to zero, then the population is multiplied by r in the next time period, and we can consider $r - 1$ as the current growth rate. If x_n becomes zero or negative, then the process has terminated with the extinction of the population. If $r < 1$, then equation (3.20) represents a declining population headed for extinction.

The maximum possible value for $x(1 - x)$ is $1/4$, which is achieved by putting $x = 1/2$. Therefore, if $r < 4$ and $x_0 < 1$, then x_1 is also less than 1, and, similarly, for all n we have $x_n < 1$; in other words, the population remains forever less than 1. If $r = 4$, then the initial value $x_0 = 0.5$ leads to an immediate extinction of the population, but other initial values lead to the interesting fluctuations shown in Figure 3.2(d). For values of r greater than 4, the population generally collapses after a few cycles.

The value of r affects the nature of population growth

The bar graphs of Figure 3.2 show how a population can fluctuate for various values of r. Note that every bar is split vertically. Each bar graph shows two different sequences with different initial values. For example, in Figure 3.2(a), the left (hollow) half of each bar represents the population with initial value $x_0 = 0.9$, and the right (solid) half represents the population with initial value $x_0 = 0.2$. In Figure 3.2(a), both populations tend rather quickly to $1/3$. This represents a stable population.

Figures 3.2(b) and (c) show that populations governed by equation (3.20) can have periodic behavior. In Figure 3.2(b), we see a situation in which years of light and heavy population alternate, and in Figure 3.2(c) the population eventually exhibits a four-year cycle. With other values for r, eight- or sixteen-year cycles are possible. The two populations start with initial conditions that are close and eventually settle into identical periodic behavior.

Figure 3.2(d) shows a very different kind of population behavior. Here the population fluctuates from year to year in a manner that seems random. Furthermore, two populations with initial values that are very close, $x_0 = 0.6000$ and $x_0 = 0.6001$, drift apart after a few years and exhibit what appears to be independent random behavior. This shows that our ability to predict a process of this kind is severely limited because an error in the fourth decimal place is great enough to make our calculations meaningless after about 15 iterations.

A process like Figure 3.2(d) is called *chaotic* because it exhibits two properties.

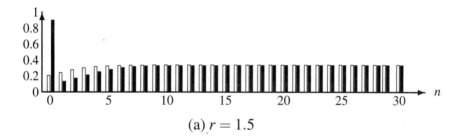

(a) $r = 1.5$

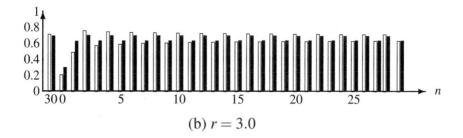

(b) $r = 3.0$

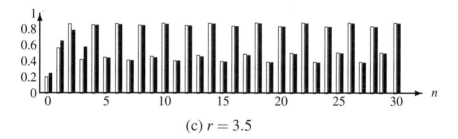

(c) $r = 3.5$

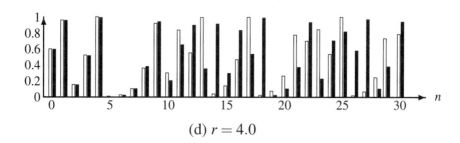

(d) $r = 4.0$

Figure 3.2. The logistic equation $x_{n+1} = rx_n(1 - x_n)$. Each of these four bar graphs shows a pair of solutions of the logistic equation with various values of r as follows: (a) $r = 1.5$, $x_0 = 0.2$ and 0.9, (b) $r = 3.0$, $x_0 = 0.2$ and 0.3, (c) $r = 3.5$, $x_0 = 0.20$ and 0.25, (d) $r = 4.0$, $x_0 = 0.6000$ and 0.6001. In (a), x_n tends to $1/3$ independent of the initial values. In (b) and (c), x_n tends toward periodic behavior with periods 2 and 4, respectively. In (d), the behavior of x_n is chaotic; although the two initial values are so close that it is impossible to distinguish them by eye in the scale shown above, the solutions soon drift far apart.

1. It fluctuates in a manner that seems random, although the process is known to be deterministic.

2. Very small changes in the initial condition, the value of x_0, produce very large discrepancies after a few cycles. The smallest error in the measurement of x_0 leads to predictions for x_n that are meaningless after a small number of cycles.

For the logistic equation (3.20), the chaotic mode starts at $r = 3.57$ and generally holds for r in the interval $3.57 \leq r \leq 4$, although there are subintervals of periodicity. In fact, for every integer, there are instances of periodicity with the given integer as a period.

This phenomenon of deterministic chaos was discovered by Edward Lorenz[2] in 1963, in the context of weather prediction. The promise of weather prediction was an important motivation in the early development of the digital computer by John von Neumann and others, but Lorenz's discovery brought about a more pessimistic attitude toward long-range weather prediction. The current thinking is that the underlying processes of weather are chaotic and that even the most precise measurements of initial conditions can never enable precise and reliable weather prediction beyond a few days.

In fact, the discovery of deterministic chaos has brought about a change in the philosophy of science. Newton's success in predicting the motion of the planets from basic scientific principles led to the optimistic attitude toward science of the Age of Enlightenment—the belief that Newton's success could be repeated in every aspect of the physical universe if only our measurements were sufficiently precise. Newton investigated a nonchaotic phenomenon, the motion of the planets, but now we know that we can never achieve the same success in predicting a chaotic process like the weather.

Our reconsideration of Zeno's paradox has led us to important facts about exponents and geometric progressions that will be very useful to us in the succeeding chapters. Our discussion of chaos takes us to the current frontiers of science.[3]

In this chapter and in the previous one we have recalled algebraic techniques sufficient to show, among other things, an interesting fact about the square root of 2 in the next chapter.

Part II

COUNTING

Chapter 4

Interesting Numbers

When you have excluded the impossible, whatever remains, however improbable, must be the truth.
—A. CONAN DOYLE, *The Adventure of the Beryl Coronet*

Mathematics has no privileged road to the truth. However, there are modes of reasoning that are seldom seen outside of mathematics. Shortly we will encounter one of these methods: *proof by contradiction*. However, this method is not as exotic as one might think; in fact, it is the mathematical equivalent of the *process of elimination*, a favorite device of Sherlock Holmes.

We will see examples of proof by contradiction. Although it has been many years since I first saw the proof, attributed to Pythagoras (c. 582– c. 500 B.C.), of the irrationality of $\sqrt{2}$, I still think that it is as astonishing as any of Sherlock Holmes' deductions.

A conversation between the English mathematician G. H. Hardy (1877–1947) and Srinivasa Ramanujan (1887–1920), the Indian mathematical genius discovered and befriended by Hardy, leads to a less serious application of proof by contradiction. When Hardy said that he arrived in a taxi with a very dull number—1729— Ramanujan responded, "No, Hardy! It is a very interesting number. It is the smallest number expressible as the sum of two cubes in two different ways."[1]

Is it possible that every natural number has some interesting special property?

Statement 4.1. *All of the natural numbers are interesting because if any of them is uninteresting, then there must be a smallest uninteresting natural number. But that is a contradiction because being the* smallest uninteresting natural number *is an interesting property!*

Common sense tells us that Statement 4.1 is a joke and not a theorem; the logical flaw is that *interesting numbers* are not well-defined. Nevertheless, we will see below that Statement 4.1 affords an introduction to proof by contradiction, an important mathematical concept. More specifically, Statement 4.1 introduces a special case of this method called *proof by infinite descent*.

Proof by Contradiction

In a direct proof we show that an assertion is true; in a proof by contradiction, also called *indirect proof* or *reductio ad absurdum*,[2] we show that it is impossible that the assertion is false—by inferring false consequences from the assumption that the assertion is false. For example, consider the assertion,

$$\text{The mailman hasn't been here today.} \tag{4.1}$$

Dr. Watson and Sherlock Holmes would each view this matter differently.

Direct proof: Dr. Watson would say, "Holmes, I have been watching for the mailman all day, and I haven't seen him."

Indirect proof: Sherlock Holmes would say, "My dear Watson, if the mailman had been here today, he would have taken your outgoing mail, which I see is still in your mailbox." Holmes sees that the denial of assertion (4.1) has implications that are inconsistent with the facts.

To prove an assertion by contradiction:

1. Assume that the assertion is false.

2. Derive false consequences from this assumption. It follows that the assumption is wrong and that the original assertion is true.

Here is an example of a mathematical proof by contradiction.

Proposition 4.2. *There do not exist natural numbers x and y such that*

$$x^2 - 4y = 3 \tag{4.2}$$

Proof. Contrary to this assertion, suppose that there do exist x and y satisfying equation (4.2). There are two cases to consider.

1. **x is even.** In this case there exists a natural number z such that $x = 2z$. Substituting this into equation (4.2), we obtain

$$(2z)^2 - 4y = 4z^2 - 4y = 3 \tag{4.3}$$

But the left side of equation (4.3) is divisible[3] by 4 and the right side is not. Contradiction!

2. **x is odd.** In this case there exists a natural number z such that $x = 2z + 1$. Again we substitute this into equation (4.2), and this time we obtain

$$(2z+1)^2 - 4y = 4z^2 + 4z + 1 - 4y = 3$$

and subtracting 1 from both sides of this equation we have

$$4z^2 + 4z - 4y = 2 \tag{4.4}$$

But, as before, the left side of equation (4.4) is divisible by 4 and the right side is not. Contradiction again!

We have shown that in every possible case equation (4.2) leads to a contradiction. We conclude that our assumption that the proposition is false leads to contradictions. Therefore, the proposition must be true. ☐

The method of infinite descent

Statement 4.1 introduces a special case of proof by contradiction called the *method of infinite descent*. The name infinite descent is from the French mathematician Pierre de Fermat (1601–65) who used this method with great success. In Statement 4.1, we use infinite descent to "prove" that there are no uninteresting natural numbers.

The method of infinite descent is a variant of the indirect method. To prove a proposition $\mathcal{P}$, we assume that $\mathcal{P}$ is false and derive the following special sort of contradiction.

1. Show that the falsehood of $\mathcal{P}$ implies that a certain nonempty set S of numbers exists. In Statement 4.1, S is the set of uninteresting natural numbers.

2. Show that there must be a smallest number n_0 in S. For Statement 4.1 there is nothing to show because *every* nonempty subset of the natural numbers has a least element.[4]

3. Show that the existence of the smallest element n_0 of S leads to a contradiction. For example, in the "proof" of Statement 4.1 the contradiction is that n_0 is *interesting* because it is the smallest *uninteresting* natural number. On the other hand, in serious mathematical applications of the method of infinite descent, the contradiction is always that there is an element of S smaller than n_0—smaller than the smallest. Showing this is the crux of a proof by infinite descent. The process is called *infinite descent* because the argument can be reapplied indefinitely to find an infinite sequence of successively smaller elements of S.

Enough of abstractions! Let us look at three examples of proof by infinite descent. In the first example, we show that it is impossible to solve a certain equation in natural numbers; the second example shows that the square root of 2 is an irrational number; and in the third example we prove a geometric theorem.

A diophantine equation

Our first example of the method of infinite descent is a diophantine equation.

Proposition 4.3. *There do not exist natural numbers x, y, and z such that*

$$x^2 + y^2 + z^2 = 2xyz$$

Proof. In fact, it is simpler to prove a somewhat stronger assertion, namely, that there do not exist natural numbers x, y, and z such that

$$x^2 + y^2 + z^2 = Pxyz \qquad (4.5)$$

> **nonempty.** A set is said to be *nonempty* if it contains *something*. All sets are nonempty except for one: the *empty set* (denoted ∅), the set that contains nothing. The empty set and the number zero have similar roles in mathematics.

> **diophantine equation.** A problem to find integer[5] solutions of an equation—after **Diophantus of Alexandria**, a third-century Greek mathematician who considered the problem of finding integer solutions to equations.

where P is a power of 2, that is, where P is one of the natural numbers $2, 4, 8, \ldots$.

We use the indirect method. Suppose there do exist solutions of equation (4.5). Then consider the nonempty set S consisting of all natural numbers of the form $x^2 + y^2 + z^2$ where x, y, and z solve equation (4.5).

The square of an even integer is even, and the square of an odd integer is odd. Moreover, the following stronger statements are true.

1. *The square of an even integer is divisible by* 4. In fact, if n is even, then for some integer k, we have $n = 2k$. It follows that $n^2 = 4k^2$, which shows that n^2 is divisible by 4.

2. *The square of an odd integer leaves a remainder of* 1 *on division by* 4. In other words, the square of an odd integer n is always of the form $n^2 = 4k + 1$ for some integer k. In fact, since an odd integer n must be of the form $n = 2h + 1$ for some integer h, squaring n we obtain

$$n^2 = (2h + 1)^2 = 4h^2 + 4h + 1 = 4(h^2 + h) + 1$$

We now see that n^2 is of the form $4k + 1$ where k is the integer $h^2 + h$.

According to equation (4.5), the sum of the three integers x^2, y^2, and z^2 is even. There are two possibilities.

1. Two of these three numbers are odd and one of them is even.

2. All three are even.

Suppose that the first alternative holds. Then one of the three numbers, x^2, y^2, and z^2 is of the form $4k$ and two of them are of the form $4h + 1$. Therefore, the sum of the three numbers is of the form $4h + 2$ for some integer h. The numbers of the form $4h + 2$ can be characterized as the integers divisible by 2 but not by 4. Dangerous curve! ↪ From equation (4.5), it follows that $Pxyz$ is divisible by 2 but not by 4. But this is impossible because P is divisible by 2, and one of the integers x, y, or z is even; therefore, the product $Pxyz$ is divisible by 4.

Since the first alternative leads to a contradiction, the second alternative must hold. That is, for every solution of equation (4.5), all three of the numbers x, y, and z are even.

Let n_0 be the smallest element of S. That is, n_0 is the smallest natural number such that for some solution x_0, y_0, and z_0 of equation (4.5) we have

$$n_0 = x_0^2 + y_0^2 + z_0^2$$

Since x_0, y_0, and z_0 are all even, there must exist integers x_1, y_1, and z_1 such that

$$x_0 = 2x_1, \ y_0 = 2y_1, \ \text{and} \ z_0 = 2z_1$$

Equation (4.5) now becomes

$$n_0 = 4x_1^2 + 4y_1^2 + 4z_1^2 = P \cdot 2x_1 \cdot 2y_1 \cdot 2z_1 \tag{4.6}$$

It is now clear that n_0 is divisible by 4. Put $n_1 = n_0/4$ and $P_1 = 2P$. Note that since P is a power of 2, P_1 is the next higher power of 2. Dividing equation (4.6) by 4, we have

$$n_1 = x_1^2 + y_1^2 + z_1^2 = P_1 x_1 y_1 z_1$$

We see that x_1, y_1, and z_1 solve equation (4.5) with P replaced by P_1. Note that n_1 is smaller than n_0. But this is impossible because we chose n_0 to be the smallest ⟵ Dangerous curve! number in the set S, and now we see that n_1 is a still smaller number in the set S. This contradiction finishes the proof by infinite descent. □

The square root of 2

For our second example of proof by infinite descent, we show the following proposition.

Proposition 4.4. *The square of a natural number cannot be equal to twice the square of another natural number.*

Note that this proposition is not disproved by the formula

$$0^2 = 2 \times 0^2$$

because, by definition, a natural number must be *positive.*

Proof of Proposition 4.4. Suppose that there are natural numbers p and q such that

$$p^2 = 2q^2 \tag{4.7}$$

This is equivalent to the assumption that Proposition 4.4 is false. If this supposition leads to a contradiction, then, according to proof by contradiction, Proposition 4.4 is true. Moreover, if Proposition 4.4 is false then there are natural numbers for which equation (4.7) is true. Of all the pairs of natural numbers (p, q) that satisfy equation (4.7), there must be a pair, which we denote (p_0, q_0), such that no pair of natural numbers (p, q) with $p < p_0$ satisfies equation (4.7). In other words, p_0 is the smallest p that solves equation (4.7).

Now observe that p_0 must be even because its square is even. Indeed, since p_0 is the smallest p occurring in a solution of equation (4.7), it must satisfy

$$p_0^2 = 2q_0^2 \tag{4.8}$$

and the right side of this equation is obviously even. We conclude that p_0 is even because if it were odd, its square would also be odd.

Furthermore, every even number is equal to two times some suitable integer. Since p_0 is even, there is an integer r_0 such that $p_0 = 2r_0$, and equation (4.7) becomes

$$(2r_0)^2 \equiv 4r_0^2 = 2q_0^2$$

and by canceling the 2's and reversing the order of r_0 and q_0, we obtain

$$q_0^2 = 2r_0^2 \qquad (4.9)$$

But notice that equation (4.9) is the same as equation (4.7), with p replaced by q_0 and q replaced by r_0. But it follows from equation (4.8) that q_0 is smaller than p_0. Therefore, we have found a still smaller solution for p in equation (4.7) than the one we supposed was smallest.

Dangerous curve! ↝

Let us review what we have done. We made the assumption that Proposition 4.4 is false. If this assumption is false, it means Proposition 4.4 is true. This assumption says that, contrary to Proposition 4.4, there are solutions—at least one—of equation (4.7). We define p_0 to be a smallest natural number p among the solutions of equation (4.7). Furthermore, we are able to construct a still smaller solution of equation (4.7). This clever construction—the key step in the proof—leads to the contradiction that there can be *no* smallest natural number p_0 that solves equation (4.7). In summary, we made the assumption that, contrary to Proposition 4.4, there are solutions to equation (4.7), and we found that this assumption leads to a contradiction. We conclude that our assumption is false, and, therefore, Proposition 4.4 must be true.   □

We used the above proof of Proposition 4.4 to illustrate the method of infinite descent. Now let's look at another proof of Proposition 4.4 that is simpler in the sense that it does not use an advanced method like infinite descent. However, before we discuss this proof, we need to introduce the concept of *function*.

When the value of one variable *depends* on the value of another variable, we say that the one is a function of the other. This meaning is also used outside of mathematics. In fact, one of the common meanings of *function* is, "Something closely related to another thing and dependent on it for its existence, value, or significance." A student's grade is a *function* of how well she does on the exams. However, in mathematics, the fact that one variable is a function of another does not imply a *causal* connection between the variables. For example, the area of a circle is a function of its radius. It is less usual, but equally correct, to say that the radius is a function of the area; however, no causal connection is implied in either case. If A is the area and r is the radius of a circle, then the variable A is a function of the variable r. In fact, we can express this dependence in a very precise way:

$$A = \pi r^2$$

On the other hand, what if we know that there is a relation between these variables, but we don't know precisely what form it takes? Then we introduce a symbol to represent the function. We use f in this example to represent the function that expresses the dependence of the area A on the radius r. We can equally well use any convenient letter; in fact, we will shortly use E to represent a different function. One can imagine that f is a machine that accepts numbers, the possible values for the radius, and gives back numbers, the corresponding values for the area. The notation $f(r)$ denotes the value that f associates with radius r. Using the variable A, previously defined to represent the area, we have

$$A = f(r)$$

Be cautioned that $f(r)$ does not represent a product of two numeric variables f and r. Despite this ambiguity, $f(r)$ is the standard notation for functions. Bear in mind that the variable f is unlike the variables that we have encountered so far. The other variables have values that are numbers, but f represents a function, a more complex mathematical object.

Now we are ready to start the proof.

Alternate proof of Proposition 4.4. We begin with a strange definition. Note that in the context of the natural numbers, "divides" means the same thing as "divides evenly." We define a function $E(n)$ that counts how many times 2 divides the natural number n. More precisely, for any natural number n, let $E(n)$ denote the largest power of 2 that divides n. In other words, $N = E(n)$ means that 2^N divides n but 2^{N+1} does not divide n. In particular, if n is odd, then $E(n) = 0$. We will make a formal definition for this sense of "divides" in a minute, but first let's finish the alternate proof of Proposition 4.4.

Every natural number can be represented as the product of a power of 2 and an odd number. For example,

$$5 = 2^0 \cdot 5, \quad 6 = 2^1 \cdot 3, \quad 56 = 2^3 \cdot 7$$

Because $E(n)$ counts the number of times 2 divides n, in each of these examples, the exponent of 2 is precisely the value returned by the function E:

$$E(5) = 0, \quad E(6) = 1, \quad E(56) = 3$$

Notice that for any natural number n, we have $E(n^2) = 2E(n)$ because 2 divides $n^2 (= n \cdot n)$ twice as many times as it divides n. To satisfy any doubt, try a few specific numbers for n. Also notice that for any natural number n, $E(2n) = E(n) + 1$ because 2 divides $2n$ once more than it divides n. Now look at equation (4.7). If there are any natural numbers p and q that satisfy equation (4.7), then we must have

$$E\left(p^2\right) = E\left(2q^2\right) \qquad (4.10)$$

But the left side of equation (4.10) is equal to $2E(p)$ and therefore must be even; on the other hand, the right side of equation (4.10) is equal to $2E(q) + 1$ and therefore must be odd. Contradiction! □

Let us return to the matter of the mathematical usage of the word *divide* in the context of the natural numbers. Here is a formal definition for *divide*.

Definition 4.5. Let n and m be natural numbers. If there exists a natural number r such that $n = mr$, then we write $m \mid n$ and we say

- *m divides n*
    or
- *m is a factor of n*
    or
- *n is divisible by m.*

Notation. If y is a function of x, and F is the name of that function, then we write $y = F(x)$, and we say, "y equals F of x." This means that the value of y depends on (is determined by) the value of x. For example, if $F(x)$ is defined to be $2x$, then

$$F(3) = 2 \times 3 = 6$$
$$F(z+3) = 2(z+3)$$
$$= 2z + 6$$

↩ Dangerous curve!

Note that this definition does not refer to the usual arithmetic operation of division.

There is another way of looking at what we have proved in Proposition 4.4. First we must review the meaning of two concepts of elementary algebra: *square root* and *rational number*.

Recall that the square root of 2 is a number x such that $x^2 = 2$. One could say *a* square root of 2 instead of *the* square root because there are two such numbers, a positive one and a negative one: $\pm 1.4142\ldots$. However, there is a convention that the *positive* square root is called *the* square root and is represented by the symbol $\sqrt{2}$.

On the other hand, the rational numbers are the fractions—proper and improper. A rational number is a number that can be expressed as the quotient of two integers: p/q.

> Q. Why do we call these numbers rational? Are the other numbers insane?
>
> A. The numbers that are not rational are called irrational. The ancient Greeks are responsible for these terms. They had a difficult time believing that there are any numbers other than the rational numbers, but they realized that mathematics is incomplete unless one goes beyond the rational numbers. They settled the matter by making a distinction between arithmetic (rational) magnitudes and geometric magnitudes.

As the reader may have guessed, we now ask whether $\sqrt{2}$ is a rational number. What does this mean? If $\sqrt{2}$ is rational, it means that there are (positive) integers p and q such that

$$\left(\frac{p}{q}\right)^2 = 2$$

which is equivalent to equation (4.7). Therefore, since we have shown that there are no natural numbers that solve equation (4.7), it must be true that $\sqrt{2}$ is not rational.

Repeating Decimals

In the previous section, we showed that $\sqrt{2}$ is irrational. The decimal expansion of this number is

$$1.41421356237309504884\ldots$$

There seems to be no repetition of the digits, but we cannot prove this simply by examining a large number of digits. Nevertheless, in this section, we will prove that the irrational numbers—$\sqrt{2}$ is one of them—can be characterized by a certain property of their decimal expansions, the property that their decimal expansions are *nonrepeating*. On the other hand, the rational numbers are precisely those numbers that have repeating decimal expansions. Let us discuss what we mean by a *repeating* decimal fraction.

Certain decimal fractions, like

$$3.3333\ldots \text{ and } 17.3012012012\ldots$$

are called *repeating decimals*. We indicate the repeating part of a decimal expansion by placing a horizontal bar over the repeating digits. For example, using this notation, the above examples of repeating decimals are written

$$3.3333\ldots = 3.\overline{3}$$

$$17.3012012012012\ldots = 17.3\overline{012}$$

> *Q. Is a whole number like 5 a repeating decimal?*
>
> *A. Yes. A number is called a repeating decimal if it has at least one representation as a repeating decimal. The number 5 is a repeating decimal because it can be written either as $5.\overline{0}$ or $4.\overline{9}$.*

This is an appropriate place in which to discuss repeating decimals because we will see that the repeating decimals are exactly the same as the rational numbers. Joke: Since these numbers repeat themselves, perhaps they can be considered uninteresting in the sense of Statement 4.1.

Here is the main result concerning repeating decimals.

Proposition 4.6. *A number is a repeating decimal if and only if it is rational.*

If and only if

Since the phrase "if and only if" occurs in the statement of many theorems and propositions, let us take a moment to clarify it. The proof of every *if and only if* proposition comes in two parts: we must show *if* and we must show *only if*. The *if* clause is called a *sufficient condition*, and the *only if* clause is called a *necessary condition*.

In stating necessary and/or sufficient conditions, the word *implies* and the symbols $\Rightarrow$ and $\Leftarrow$ are often used.

The following have exactly the same meaning:

- If *Socrates is a man*, then *Socrates is mortal*.

- *Socrates is a man* only if *Socrates is mortal*.

- That *Socrates is a man* is a sufficient condition that *Socrates is mortal*.

- That *Socrates is mortal* is a necessary condition that *Socrates is a man*.

- *Socrates is a man* implies that *Socrates is mortal*.

- *Socrates is a man* $\Rightarrow$ *Socrates is mortal*.

- *Socrates is mortal* $\Leftarrow$ *Socrates is a man*.

Many mathematical propositions are stated in this fashion: "If something is true, then something else is true." In the model proposition,

If *Socrates is a man*, then *Socrates is mortal*.

"*Socrates is a man*" is called the *hypothesis*, and "*Socrates is mortal*" is called the *conclusion* of the proposition. We have listed above seven different ways of stating this proposition.

There are alternate ways of stating an "if and only if" proposition as follows:

- *Socrates is mortal* if and only if *Socrates will not live forever*.

- That *Socrates is mortal* is a necessary and sufficient condition that *Socrates will not live forever*.

- *Socrates is mortal* ⇔ *Socrates will not live forever*.

The word *characterize* has a special meaning in mathematics—it gives still another way of stating necessary and sufficient conditions. For example, the fourth sentence of this section says, in part, "the irrational numbers ... can be *characterized* by a certain property of their decimal expansions." This is another way of stating that a number is irrational *if and only if* its decimal expansion has a "certain property." We refer, of course, to the property of having a nonrepeating decimal expansion. In contrast, the nonmathematical usage of *characterize* does not carry the meaning of *if and only if*. For example, in ordinary usage, if we say, "Ada *characterized* the speakers as arrogant bores," we mean that Ada claims that all of the speakers are arrogant bores. On the other hand, if we (perversely) insist on the mathematical usage, then this sentence means that all of the speakers *and no others* are arrogant bores.

Proving Proposition 4.6

algorithm. A completely specified step-by-step computational procedure. A "mathematical recipe."

Now let us return to the proof of Proposition 4.6. The algorithm of long division gives us a way of finding as many digits as we please of the decimal expansion of a quotient. Therefore, it is clear that every rational number has at least one representation as an infinite decimal fraction. It is a troubling fact that some rational numbers have two representations as infinite decimals. For example, we have,

$$1.200000\cdots = 1.199999\ldots \tag{4.11}$$

Before we go further, let us establish a convention that excludes a decimal expansion that ends in an infinite sequence of 9's, like $1.1999999\ldots$, because there is always an alternate representation, such as we see in equation (4.11), *without* an infinite sequence of 9's. This convention makes the decimal representations unique. Notice that long division always gives us the representation that conforms with this convention. We use this convention in the proof below of Proposition 4.6.

Proof (if). In the *if* part of the proof we must show that every rational number can be expressed as a repeating decimal fraction. This follows from looking at the process of *long division*, which generates the digits of the decimal expansion. The digits of the decimal expansion *from some point on* are completely determined by the current remainder.

> *Q. It seems to me that the digits are always determined by the re-*
> *mainders. Why do you say "from some point on"?*
>
> *A. The situation is a bit more complicated. At some point in the*
> *long division algorithm, we "bring down" only zeros to append to the*
> *remainder. Therefore, it is only after this point that the current remainder*
> *determines, without reference to the digits of the quotient, (1) the next*
> *digit of the quotient and (2) the next remainder, and so on.*

Look at the remainders that occur after we reach the point in the long division algorithm we just discussed. Only finitely many remainders are possible. In fact, for the rational number of the form p/q, only q different remainders are possible: $0, 1, \ldots, q - 1$. When we encounter the same remainder twice, then, since we "bring down" a zero in both instances, the sequence of digits of the quotient must repeat forever after.[6]   □

Proof (only if). In this part of the proof, we show that if a number is a repeating decimal, then it is rational. We do not show this in general because a special case demonstrates the general method quite well. We will show that the assertion holds for the repeating decimal $17.3\overline{012}$, which we call R.

We could prove that R, that is, $17.3\overline{012}$, is rational by using Theorem 3.8 (the sum of a geometric progression). However, it is simpler to proceed as follows. Observe that $1000 \times R$ is equal to $17301.2\overline{012}$. Hence we have:

$$1000R - R = 17301.2\overline{012} - 17.3\overline{012} = 17301.2 - 17.3 = 17283.9 \qquad (4.12)$$

But notice that the left side of equation (4.12) is equal to $999R$. Dividing both sides of equation (4.12) by 999 gives

$$R = \frac{17283.9}{999} = \frac{172839}{9990}$$

Since we have expressed R as a quotient of two integers, we have succeeded in showing that R is rational.   □

Lines and Points

A geometric application of infinite descent

Our third proof by infinite descent concerns a finite set of points in two- or three-dimensional space.

Proposition 4.7. *Let S be a finite set of points. Suppose that a straight line connecting any two of the points of S always contains a third point of S. Then all of the points of S lie on a single straight line.*

This assertion seems plausible, but its proof is rather subtle.

Proof. We proceed by the indirect method. Suppose that the points in the finite set S do not all belong to the same straight line. Then at least one point of S is at a *positive* distance from a line determined by two other points in S. Since there must be finitely many such positive distances, one of them, d_0, must be the *smallest*. Suppose that this smallest positive distance is realized, as shown in Figure 4.1(a), by the point P in S and the line l_0 that contains points A and B, which also belong to S. By assumption, there is a third point C (possibly among others) on the line l_0 that also belongs to S.

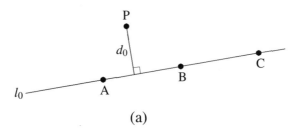

(a)

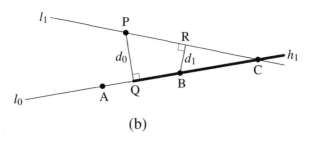

(b)

Figure 4.1. Proposition 4.7.

In Figure 4.1(b), we define Q to be the point on l_0 (not necessarily in S) such that PQ is perpendicular to l_0. The distance $\overline{PQ}$ is equal to d_0, the distance between point P and line l_0. Either the right or left half-line emanating from Q—we denote Dangerous curve! ↬ it h_1—must contain at least two of the three points A, B, and C.[7] In Figure 4.1(b), h_1 is the right half-line emanating from Q because B and C are on the right side of Q. Of the points in S on h_1—there are at least two of them—one is farthest from Q, and a second one of them is nearest to Q; call these points the *far* point and the *near* point, respectively. The near point is between the far point and Q; the near point may coincide with Q, but not with the far point. In Figure 4.1(b), C is the far point and B is the near point. Define l_1 to be the line through P and the far point (C). Consider the distance d_1 between the near point (B) and the line l_1.

From Figure 4.1(b), it seems clear that the distance d_1 is less than the distance d_0. But this contradicts the assumption that d_0 is the shortest distance between any point in S and any line determined by two points of S. The proof is complete apart from a rigorous demonstration of the plausible fact that d_1 is less than d_0.[8]   □

Desargues's theorem

Proposition 4.7 is a geometric result that has a different flavor from the familiar geometry taught in school because it deals only with the incidence of lines and points and avoids any mention of distances or angles.[9] The following is a famous and important geometric result concerning the incidence of lines and points in three-dimensional space that is named after its discoverer, the French architect and geometer Gérard Desargues (1593–1662).

Theorem 4.8 (Desargues's theorem). *In Figure 4.2, the triangles* ABC *and* A′B′C′ *are shown. The points* A *and* A′ *are arbitrarily designated corresponding points; similarly,* B, B′ *and* C, C′ *are pairs of corresponding points. The sides* AB *and* A′B′ *are called* corresponding sides; *similarly,* BC, B′C′ *and* CA, C′A′ *are pairs of corresponding sides. If the lines determined by the three pairs of corresponding points are concurrent (i.e., if they meet in a point denoted* O), *then the straight line extensions of corresponding sides meet in collinear points (i.e., the intersection points* X, Y, *and* Z *lie on a straight line).*

We prove this theorem under the following *additional assumptions.*

1. Triangles ABC and A′B′C′ do not lie in parallel planes, and no pair of corresponding sides of these triangles are parallel.

2. Triangles ABC and A′B′C′ are not coplanar.

Proof. Each pair of corresponding sides of the two triangles ABC and A′B′C′ is contained in a plane; for example, AB and A′B′ are contained in the plane of the triangle ABO. Therefore, pairs of corresponding sides, since they are coplanar and not parallel, must intersect; we denote the intersection points X, Y, and Z. The points X, Y, and Z must must belong to the intersection of the planes of ABC and A′B′C′. Since the intersection of nonparallel planes is a straight line, we see that, as claimed, the points X, Y, and Z are collinear.   □

incidence. In geometry, a line and point are said to be *incident* if the point lies on the line. Euclidean geometry, also called *metric* geometry, deals not only with the incidence of lines and points, but also with distances and angles.

coplanar. Lines are said to be coplanar if they lie in a single plane.

collinear. Points are said to be collinear if they lie on a single line.

Projective geometry was developed by the French mathematician Jean-Victor Poncelet (1788–1867) during the period 1813–14 while he was a prisoner of war in Russia. Poncelet was an officer in Napoleon's Grand Army, a force of 500,000 that suffered near annihilation in 1812 during the retreat from Moscow.

Desargues's theorem is of fundamental importance in *projective geometry*, an investigation inspired in part by the technique of linear perspective in art. It may seem surprising that the ancient Greek geometers overlooked Desargues's theorem, but a possible reason is that the system of linear perspective in drawing and painting was developed at a much later time in the Renaissance. The principles of linear perspective were codified by the Florentine architect Filippo Brunelleschi in the period 1417–20. In linear perspective, three-dimensional parallel lines are depicted as converging lines in a two-dimensional drawing or painting. The intersection of these converging lines is called a *vanishing point*, a concept related to *point at infinity* in projective geometry. Linear perspective is shown very dramatically in Leonardo da

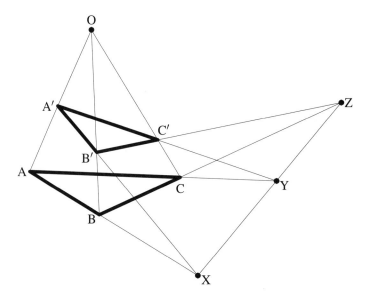

Figure 4.2. Desargues's theorem.

Vinci's celebrated fresco, *The Last Supper*, in which there is a vanishing point at Christ's head. There can be more than one vanishing point. The vanishing point of every system of parallel *horizontal* lines lies on a straight line called the *horizon line*, a concept related to *line at infinity* in projective geometry.

In projective geometry, lines and points at infinity are called *ideal elements*; they are introduced as follows. We augment each plane by introducing one additional line, a *line at infinity*; parallel planes share the same line at infinity. We augment each line with one additional point, a *point at infinity*; parallel lines share the same point at infinity. If a line lies in a plane, then the given line's point at infinity is contained in the given plane's line at infinity. These incidence properties of the ideal elements do not need proof because they are part of the axiomatic structure of projective geometry—made plausible by the concepts of vanishing point and horizon line in perspective drawing. On introduction of the ideal elements, the following **Axioms of Incidence** are true in three-dimensional projective geometry.

1. Every pair of points is contained in exactly one line.

2. Every pair of planes intersects in exactly one line.

3. Every pair of coplanar lines intersects in exactly one point.

Note that in Euclidean geometry, the Axioms of Incidence must make exceptions in the case of parallel lines or planes. The introduction of ideal elements enables us to avoid mention of these special cases.

Assumptions 1 and 2 of Desargues's theorem can be removed

Assumption 1 of Theorem 4.8 is removed by introducing ideal elements. The above proof holds without change with the understanding that any of the lines and points might be ideal elements. Parallelism of lines or planes does not play a role in projective geometry.

Assumption 2 of Theorem 4.8—that the two triangles ABC and A'B'C' are not in ↤ Dangerous curve! the same plane—can be eliminated by using the bright idea that Figure 4.2 is a two-dimensional diagram that is a projection of a three-dimensional configuration. We can imagine that, under illumination from the sun, Figure 4.2 is the two-dimensional shadow on a flat surface of a three-dimensional configuration. The fact that in three dimensions the points X, Y, and Z are collinear implies that the two-dimensional shadows of these three points are also collinear. The two-dimensional version of Desargues's theorem follows from the fact that any two-dimensional figure satisfying the hypotheses of Theorem 4.8 is the projection of at least one three-dimensional configuration.

Desargues's configuration

It is interesting to note that Figure 4.2 consists of ten points

$$A, B, C, A', B', C', O, X, Y, Z$$

and ten lines such that:

1. Every line contains exactly three of the points.

2. Every point is the intersection of exactly three of the lines.

A set of n points and n lines with properties 1 and 2 is called a *configuration*. For example, Figure 4.2 is called *Desargues's configuration*.

The principal results of this chapter, Propositions 4.4 and 4.6, come together to show that $\sqrt{2}$ has a nonrepeating decimal expansion. Along the way to this result we discussed many important mathematical concepts: proof by contradiction, infinite descent, well-ordered sets, functions, divisibility, rational numbers, and necessary and sufficient conditions. An example of proof by infinite descent led us to a glimpse of projective geometry.

In the next chapter, we see systematic methods of counting applied to the game of tenpins.

Chapter 5

Tenpins, and Counting

> *Then at the high feast evermore they should be fulfilled the whole number of an*
> *hundred and fifty, for then was the Round Table fully complished.*
> —THOMAS MALORY (c. 1400–71), *Le Morte d'Arthur*

It is not hard to imagine that the Knights of the Round Table drank a toast at their high feast. The following problem deals with an unlikely sequel to the toast.

Problem 5.1. If each knight touched glasses with every other one of the 150, how many clinks were heard?

In this chapter we will find a connection between this problem and the following seemingly disparate matters.

- In the sixth century B.C., the Pythagoreans discovered how to count the number of objects in triangular arrays of arbitrary size.

- In his youth, Isaac Newton (1642–1727) discovered the *binomial theorem*, a method of expanding certain algebraic expressions.

Triangular Numbers

In the game of tenpins, bowling pins are set up in a triangular pattern. Let us generalize this pattern as shown in Figure 5.1.

We can continue this sequence of numbers

$$1, 3, 6, 10, 15, 21, 28, \ldots$$

as far as desired. These numbers, which were supposed by the Pythagoreans to have mystical properties, are called *triangular numbers*. The nth triangular number, where n is an arbitrary natural number, is denoted T_n. Note that T_n is the sum of the first

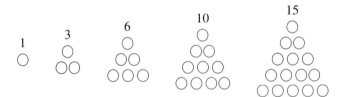

Figure 5.1. Triangular bowling pin arrangements.

n natural numbers. Is there a formula that gives T_n for all natural numbers n? After some very clever guesswork, or by looking it up in a book, we might find the formula

$$T_n = \frac{n(n+1)}{2} \tag{5.1}$$

It is easy to verify that this formula is correct for all of the cases shown in Figure 5.1.

> *Q. If what you say is true, then $n(n+1)/2$ must always be a whole number. Is this obvious?*
>
> *A. Since the natural numbers n and $n+1$ are consecutive, one of them must be even. The product of an odd number and an even number must be even. This means that the product $n(n+1)$ must be even; and therefore, the result of dividing that number by 2 must be a whole number. This shows that the right side of equation (5.1) must be a whole number.*

There is a method of proof that is particularly suitable for demonstrating a formula such as equation (5.1). It is called the method of *mathematical induction*; we will also call it the *domino theory* after the belief held early in the Vietnam War that if one Southeast Asian country fell to the Communists, then all the countries would fall. This theory was a failure in Southeast Asia, but we will have success in applying it to the triangular numbers.

Solution by mathematical induction

Let us state mathematical induction—the domino theory—in general terms because it is useful for many other problems. Suppose that we have a sequence of assertions, call them

$$\mathcal{A}_1, \mathcal{A}_2, \mathcal{A}_3, \ldots \tag{5.2}$$

The assertion corresponding to the natural number n is $\mathcal{A}_n$. We can prove the entire sequence of assertions by carrying out the following two steps.

1. Show that $\mathcal{A}_1$ is true.

2. Show that if any assertion in the series (5.2) is true, then its neighbor to the right is also true. In other words, assume that for some particular natural number n greater than 1, $\mathcal{A}_{n-1}$ is true. (This assumption is called the *hypothesis of induction*.) Then use this assumption to show that, for this same fixed n, $\mathcal{A}_n$ is also true.

In other words, we must show that if any assertion $\mathcal{A}_{n-1}$ is true, then its successor $\mathcal{A}_n$ is also true.

The assertions are like dominos.

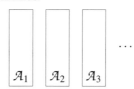

All the dominos fall if the following two things are true.

1. The $\mathcal{A}_1$ domino falls.

2. If any domino falls, then its neighbor to the right also falls.

Recall that we have defined T_n to be the sum of the natural numbers 1 through n. The following fact simply states that we get the same sum if we add all the natural numbers *less* than n (this is T_{n-1}), and then add n.

Fact 5.2. *For any natural number n,*

$$T_n = T_{n-1} + n$$

Let's state the formula for the triangular numbers in the form of a proposition.

Proposition 5.3. *Let n be a natural number. The triangular number T_n satisfies equation (5.1).*

Proof. To apply the domino theory, we need to see that the proposition is equivalent to a sequence of assertions like (5.2). We obtain such a sequence of assertions by substituting the natural numbers $1, 2, 3, \ldots$ in equation (5.1). In particular, put

$$\mathcal{A}_1 : \qquad\qquad T_1 = \frac{1(1+1)}{2} \qquad\qquad (5.3)$$

$$\mathcal{A}_2 : \qquad\qquad T_2 = \frac{2(2+1)}{2}$$

$$\mathcal{A}_3 : \qquad\qquad T_3 = \frac{3(3+1)}{2}$$

$$\mathcal{A}_n : \qquad\qquad T_n = \frac{n(n+1)}{2}$$

Q. I thought that an assertion must be a complete sentence with a subject and a verb. Are formulas like equation (5.3) considered assertions?

A. Yes. For the equal sign in equation (5.3) we read, "is equal to."

Now we must carry out the two steps.

1. Assertion $\mathcal{A}_1$ is true because the right side of equation (5.3) is equal to 1, and we know that the first triangular number T_1 is equal to 1.

2. We must show the following. If for some natural number n, $\mathcal{A}_{n-1}$ is true, then $\mathcal{A}_n$ is also true for this same choice of n. To see what $\mathcal{A}_{n-1}$ means, we must replace n with $n-1$ in equation (5.1), which gives

$$T_{n-1} = \frac{(n-1)\left((n-1)+1\right)}{2}$$

which by an algebraic simplification is

$$= \frac{(n-1)n}{2} \tag{5.4}$$

To carry out step 2, we use the assumption that equation (5.4) is true to show that $\mathcal{A}_n$ is true. We do that as follows. Fact 5.2 asserts

$$T_n = T_{n-1} + n$$

and by equation (5.4) we have

$$T_n = \frac{(n-1)n}{2} + n \tag{5.5}$$

We are finished if by algebraic manipulation we can show that the right side of equation (5.5) is equal to the right side of equation (5.1). We leave that small technical detail as an algebraic exercise for the reader. Apart from this detail, the proof is complete. $\square$

Even if we carry out this small algebraic detail, we may still feel that we don't really understand *why* formula (5.1) works. One can understand every step of a proof and still feel that something important is missing. In particular, verifying every step of the proof of a proposition—especially a proof by mathematical induction—may not reveal a natural way of discovering the proof. We discovered equation (5.1) either by trial and error or looking it up in a book. Our misgivings can be laid to rest by the following more heuristic geometric method of proving equation (5.1).

Geometric solution

Figure 5.2 shows a way of understanding equation (5.1). Notice that the small squares in Figure 5.2 are arranged almost like the circles in Figure 5.1. The difference is that in Figure 5.1 the circles are arranged in the shape of an equilateral triangle, but in Figure 5.2 the squares are arranged in the shape of a right triangle.

As the reader doubtless knows, the area of a square of side s is s^2. In Figure 5.2, the length of the sides of the small squares is 1 unit; therefore, the area of the square ABDC is $5 \times 5 = 25$. In fact, by filling in some missing squares we can actually count 25 small squares, and the area of each of them is equal to 1. In the version of Figure 5.2 corresponding to T_n, the square would have side n, and the area would be n^2. Since the right triangle ADC is half of the square, it follows that its area is $5^2/2$. This area is less than T_5; in fact, it is less by the areas of the five small triangles above the diagonal of the square. Adding the areas of those five triangles, each with area $1/2$, gives us an area of

$$\frac{5^2}{2} + \frac{5}{2} = 15$$

which is equal to T_5.

It may seem more direct simply to count the 15 squares in Figure 5.2, but the advantage of the longer method is that it generalizes easily. Extending to the n by n case, the area of the large triangle is $n^2/2$. To get T_n, we must add n small triangles above the diagonal, which gives

$$T_n = \frac{n^2}{2} + \frac{n}{2}$$

which is equivalent to equation (5.1).

There is a curious aspect to this proof. The problem of evaluating the triangular numbers has nothing to do with fractions; yet the fraction $1/2$ appears in this geometric proof. This proof violates the philosophical principle known as *Ockham's razor* that is from the scholastic philosopher William of Ockham (1285?–1349?). *Essentia non multiplicanda sunt praeter necessitatem,* which means, *Avoid introducing*

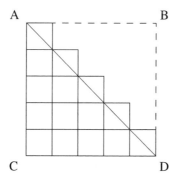

Figure 5.2. The triangular number T_5.

unnecessary concepts. At first glance, fractions have nothing to do with our problem; yet by introducing fractions we find a solution that displays the essence of the problem. Moreover, our problem doesn't have anything to do with geometry, even though we used geometry in the proof.

It is possible to give a version of this geometric argument that does not make use of fractions. However, it often happens in mathematics that the solution of a problem involves concepts that do not occur in the statement of the problem. In this case one might ask for a "more elementary" or "more natural" proof that avoids the extraneous concept. The "more elementary" proof sometimes turns out to be more difficult, as in the "elementary" proof of the prime number theorem.[1]

Algebraic solution: counting pairs

There is another proof of the formula (5.1) for the nth triangular number that is more in accord with Ockham's razor. This proof may make us feel we were mistaken in thinking that in the previous geometric proof we found the "essence of the problem." Perhaps we are like the blind men in the Indian folktale each of whom thinks that he has found the true essence of the elephant. In truth, one man felt only the leg, another only the tusk, another only the side, and another only the tail.

This new discussion of triangular numbers proceeds by counting the number of pairs of objects taken from a finite set. More concretely, consider the troubles of an imaginary firm, called *Chatterbox Inc.*, in which every employee must have a 5-minute phone conversation with every other employee at least once every week. Chatterbox Inc. does not benefit from economies of scale. In fact, if Chatterbox Inc. has sufficiently many employees, they do not have time to do anything except call each other on the telephone. To the extent that collaborators behave in this manner, this example shows the advantage of limiting the size of a collaborative team. The question is this: If Chatterbox Inc. has n employees, how many phone calls are made each week?

This problem relates to a difficulty in managing a large software project. A complex software program suffers from the possibility of unwanted interactions—called side effects—between various elements of the program. The calculation we are about to make shows how the number of possible interactions grows as the number of elements of the program grows. Many advanced programming languages have elaborate constructs to insulate various parts of the program. These languages—especially object-oriented languages such as C++ and SmallTalk—enable the programmer to permit interactions only in a very controlled manner.

Let us return to Chatterbox Inc. and the problem of counting phone calls. Since it does not matter who originates a call, the number we seek is the number of ways of choosing 2 distinct items from a collection of n distinct items; the mathematical symbol for this number is $\binom{n}{2}$.

We will find $\binom{n}{2}$ by two different methods. The first method shows a connection with triangular numbers. Let us say that the firm is made up of six people: Ada, Ben, Cal, Dot, Eli, and Fay. We make a list of all of the conversations by first listing all of Ada's conversations. Then we list all of Ben's conversations except the

The symbol $\binom{n}{2}$ is read *n choose* 2.

The symbol $\binom{n}{r}$, *n choose r*, represents the number of ways of choosing r items from n distinct items. To say that items are *distinct* means that they are all different. On the other hand, if some of the items are indistinguishable one from another, then the number of ways of choosing r distinct items is smaller than $\binom{n}{r}$.

conversation with Ada which is already counted on Ada's list. Then we list all of Cal's conversations, except the conversations with Ada and Ben because they have already been counted on the lists of Ada and Ben—and so on.

> Ada has 5 conversations with Ben & Cal & Dot & Eli & Fay.
> Ben has 4 conversations with   Cal & Dot & Eli & Fay.
> Cal has 3 conversations with   Dot & Eli & Fay.
> Dot has 2 conversations with   Eli & Fay.
> Eli has 1 conversation with   Fay.
>   15 is the total number of conversations.

There is no row of Fay's conversations, because all of them have already been listed. This shows that the total number of conversations is 15. Without even counting, it is clear from the way that this table is arranged that the number of conversations, which is denoted $\binom{6}{2}$, is equal to T_5. In general, by extending this table we see

$$\binom{n}{2} = T_{n-1}$$

Since we have already proved formula (5.1), we are able to write down a formula for $\binom{n}{2}$. However, we prefer to derive the formula for $\binom{n}{2}$ independently. This gives yet another proof of equation (5.1). In fact, the proof is very simple indeed. For each of the n people, there are $n-1$ other people who must be contacted, making a total of $n(n-1)$ conversations. But in this enumeration, every conversation is counted twice. Counting this way, the conversation between Ada and Ben is counted both in the list of Ada's conversations and in the list of Ben's conversations. Therefore, we obtain the formula

$$\binom{n}{2} = \frac{n(n-1)}{2}$$

Now we can apply this formula to solve Problem 5.1. There were 150 knights. Therefore, if every knight touched every other knight's glass, the number of clinks was

$$\binom{150}{2} = \frac{150 \times 149}{2} = 11,175$$

  Q. I think that we can conclude that this is not what happened at the Round Table. Are there similar formulas for $\binom{n}{3}$ and so on?

  A. Yes. As a matter of fact, as we will see, a similar argument shows

$$\binom{n}{3} = \frac{n(n-1)(n-2)}{1 \cdot 2 \cdot 3}$$

$$\binom{n}{4} = \frac{n(n-1)(n-2)(n-3)}{1 \cdot 2 \cdot 3 \cdot 4}$$

Binomials

Now let us examine yet another interpretation of the triangular numbers that relates to a discovery made by Isaac Newton in his youth. The following example uses some algebraic calculations. An algebraic expression with two terms like $x + y$ is called a *binomial*. Let us multiply two binomials:

$$(A + B)(a + b)$$

First multiply $a + b$ by A, obtaining $Aa + Ab$. Then multiply $a + b$ by B, obtaining $Ba + Bb$. Adding these two results we obtain

$$(A + B)(a + b) = Aa + Ab + Ba + Bb$$

And now, for a second example, let us compute

$$(A + B)(a + b)(\alpha + \beta)$$

We multiply the previous result first by α and then by β and add the two results, obtaining

$$\begin{aligned}(A + B)(a + b)(\alpha + \beta) \\ = Aa\alpha + Ab\alpha + Ba\alpha + Bb\alpha \\ + Aa\beta + Ab\beta + Ba\beta + Bb\beta \quad (5.6)\end{aligned}$$

We do not need to continue with even more complicated multiplications. The point is that the right side of (5.6) is equal to the sum of all possible products of three factors where the first factor is A or B, the second factor is a or b, and the third factor is α or β.

Now suppose that we multiply three identical binomial factors like

$$(x + 1)(x + 1)(x + 1)$$

which we could also write as $(x + 1)^3$. We can expand this expression using equation (5.6) by making suitable substitutions. In fact, putting A, a, and α equal to x; and B, b, and β equal to 1 in equation (5.6) we obtain

$$\begin{aligned}(x + 1)(x + 1)(x + 1) \\ = (x \cdot x \cdot x) + (x \cdot 1 \cdot x) + (1 \cdot x \cdot x) + (1 \cdot 1 \cdot x) \\ + (x \cdot x \cdot 1) + (x \cdot 1 \cdot 1) + (1 \cdot x \cdot 1) + (1 \cdot 1 \cdot 1) \quad (5.7)\end{aligned}$$

Some of the terms on the right side of equation (5.7) are equal. For example, the second, third, and fifth terms

$$x \cdot 1 \cdot x, 1 \cdot x \cdot x, \text{and } x \cdot x \cdot 1$$

are all equal to x^2. The expansion (5.7) is the sum of eight products. If we call one of these products abc, then

---

Notation. α and β are the lowercase Greek letters alpha and beta, respectively. Greek letters are used freely in mathematics for the names of variables. Frequently, as here, α and β are used to represent variables that have some relationship to a and b. The capital Greek letters alpha and beta are indistinguishable from the Roman letters A and B.

factor. Recall that numbers or algebraic expressions that are multiplied to form a product are called *factors* of the product.

expansion. A more verbose form of an algebraic expression is called an *expansion* of that expression. The right side of equation (5.7) is an expansion of the left side.

> *a* must be either *x* or 1,
>
> *b* must be either *x* or 1,
>
> and
>
> *c* must be either *x* or 1.

The number of terms equal to x^2 *is the number of ways of choosing two x's and one* 1. This is the number 3 *choose* 2, denoted $\binom{3}{2}$, which is equal to 3. This jibes exactly with our observation that there are three terms of the expansion of $(x+1)^3$ that are equal to x^2. And, indeed, this pattern holds in general. For any natural number *n*, the number of terms equal to x^2 in the expansion of $(x+1)^n$ is $\binom{n}{2}$, which is equal to $n(n-1)/2$. In fact, according to Newton's binomial theorem the entire expansion of $(x+1)^n$ is

$$\binom{n}{0} + \binom{n}{1}x + \binom{n}{2}x^2 + \cdots + \binom{n}{n}x^n \tag{5.8}$$

For that reason $\binom{n}{r}$, which we call *n choose r*, is also called *the binomial coefficient n over r.*

Combinations and Permutations[2]

Counting problems can easily take us beyond the realm of triangular numbers. We begin with the problem of making a composite sketch of a crime suspect.

Problem 5.4. In discussion with a police artist, we have narrowed down our recollection of the facial features of the suspect to the following: 3 mouths, 4 noses, 3 pairs of eyes, 2 pairs of ears. How many different drawings can be assembled from these choices?

Solution. Having chosen any one of the 3 mouths, we can choose any of the 4 noses; thus, for mouths and noses there are $3 \cdot 4 = 12$ possibilities. Altogether, there are $3 \cdot 4 \cdot 3 \cdot 2 = 72$ faces.

The above problem illustrates the multiplicative method that applies in more complex problems.

Problem 5.5. In how many different ways can 12 people be seated at a table with 12 places?

Solution. We put any one of the 12 at the first place. Having done so, there are 11 different people who can be seated at the second place. So far, there are $12 \cdot 11 = 132$ possibilities. Continuing, we find that the total number of seating arrangements is

$$12 \cdot 11 \cdot 10 \cdot 9 \cdot 8 \cdot 7 \cdot 6 \cdot 5 \cdot 4 \cdot 3 \cdot 2 \cdot 1 = 479,001,600$$

Note that we count it as a different seating arrangement if, for example, everyone moves one place to the left.

Since products like this occur often in counting problems, we have a special notation.

Definition 5.6. The product $1 \cdot 2 \cdot 3 \cdots n$, denoted $n!$, is called *n-factorial*.

Problem 5.7. In how may ways is it possible to choose a president, vice-president, secretary, and treasurer from a club with 26 members?

Solution. The answer is $26 \cdot 25 \cdot 24 \cdot 23 = 358,800$, which can also be written $26!/22!$. This is also called the number of *permutations* of 26 things taken 4 at a time and written $(26)_4$.

Fact 5.8. *The number of permutations of n things taken r at a time is equal to*

$$(n)_r = \frac{n!}{(n-r)!}$$

Problem 5.9. In how many ways can the club of Problem 5.7 send 4 delegates to the national meeting?

Solution. Like Problem 5.7, this problem also deals with choosing 4 people from a group of 26; however, the previous answer, $(26)_4$, is too large because in the current problem the order of selection is not important. If we choose 4 people for a delegation, there are $4!$ ways of naming these people as officers. Hence, the number $(26)_4$ is too large by the factor $4! = 24$. The correct number is

$$\frac{(26)_4}{4!}$$

In the terminology of the preceding two sections, this quantity is 26 *choose* 4 which we write $\binom{26}{4}$. The numerical value is

$$\binom{26}{4} = \frac{26 \cdot 25 \cdot 24 \cdot 23}{1 \cdot 2 \cdot 3 \cdot 4} = 14,950$$

Using the method of Problem 5.7, we have the following formula for *n choose r*, which is also known as the number of *combinations* of *n* things taken *r* at a time.

$$\binom{n}{r} = \frac{n!}{r!(n-r)!} \qquad (5.9)$$

Problem 5.10. The club of Problem 5.7 must send 4 delegates to the national meeting exactly 2 of whom must be officers. How many different ways can they choose the 4 officers and the 4 delegates?

Solution. For each of the $\binom{26}{4}$ ways of choosing 4 people to be a delegation, we can choose 2 of them as officers in $\binom{4}{2}$ ways. We can choose the remaining 2 officers from the remaining 22 nondelegates in $\binom{22}{2}$ different ways. Finally, we can assign particular offices to the officeholders in $4!$ different ways. Altogether, the number of possible selections is the product of these numbers:

$$\binom{26}{4}\binom{4}{2}\binom{22}{2}4! = 14,950 \cdot 6 \cdot 231 \cdot 24 = 497,296,800$$

Pascal's triangle

Blaise Pascal (1623–62) discovered a pattern of the binomial coefficients. He arranged these numbers in a triangular pattern as shown in Figure 5.3.

Pascal noticed that each number in the interior of the triangle is equal to the sum of the two numbers immediately above it. This fact can be proved by the method of mathematical induction.

$$\binom{1}{0}=1 \qquad \binom{1}{1}=1$$
$$\binom{2}{0}=1 \qquad \binom{2}{1}=2 \qquad \binom{2}{2}=1$$
$$\binom{3}{0}=1 \qquad \binom{3}{1}=3 \qquad \binom{3}{2}=3 \qquad \binom{3}{3}=1$$
$$\binom{4}{0}=1 \qquad \binom{4}{1}=4 \qquad \binom{4}{2}=6 \qquad \binom{4}{3}=4 \qquad \binom{4}{4}=1$$
$$\binom{5}{0}=1 \qquad \binom{5}{1}=5 \qquad \binom{5}{2}=10 \qquad \binom{5}{3}=10 \qquad \binom{5}{4}=5 \qquad \binom{5}{5}=1$$

Figure 5.3. Pascal's triangle.

Polynomials

We saw that the numbers *n choose r* appear on raising the binomial $1+x$ to the *n*th power. The expression $(1+x)^n$ is a special case of a *polynomial*.

Definition 5.11. Let $a_0, a_1, \ldots, a_{n-1}, a_n$ be constants. Then the expression

$$p(x) = a_n x^n + a_{n-1} x^{n-1} + \cdots + a_1 x + a_0$$

is called a *polynomial in the variable x*. For each integer $i, (0 \leq i \leq n)$, a_i is called the *coefficient of x^i*.

The following are examples of polynomials.

$$5x^3 - 4x + 7 \qquad (2x^2 + 3)(x+1)$$

In the second example, if we wish to know the coefficients, we must compute them as follows.

$$(2x^2 + 3)(x+1) = (2x^2+3)x + (2x^2+3) = 2x^3 + 2x^2 + 3x + 3$$

To add two polynomials in x, add the corresponding coefficients:

$$(2x^3 + 2x^2 + 3x + 3) + (x^2 - 4x + 2) = 2x^3 + 3x^2 - x + 5$$

The algorithm for multiplying polynomials is a bit more complex, but it is much like ordinary multiplication. Here is the computation of $(x+3)(x^2+3x-2)$.

$$
\begin{array}{rrrrrr}
 & x^2 & + & 3x & - & 2 \\
\times & & & x & + & 3 \\
\hline
 & 3x^2 & + & 9x & - & 6 \\
x^3 & + & 3x^2 & - & 2x & \\
\hline
x^3 & + & 6x^2 & + & 7x & - & 6 \\
\end{array}
$$

This small polynomial multiplication is easy to do by hand, but a larger problem could be exceedingly tedious and subject to error. However, today computer programs exist—Maple and Mathematica are two such programs—that do algebraic problems with lightning speed and complete accuracy on desktop computers. Here are two examples of such calculations. We will shortly make use of these formulas.

$$(1+x^5)^5(1+x^{10})^4(1+x^{25})^3 = x^{140}+5x^{135}+14x^{130}+30x^{125}+51x^{120}$$
$$+74x^{115}+99x^{110}+126x^{105}+161x^{100}+204x^{95}+246x^{90}$$
$$+281x^{85}+299x^{80}+304x^{75}+306x^{70}+304x^{65}+299x^{60}$$
$$+281x^{55}+246x^{50}+204x^{45}+161x^{40}+126x^{35}+99x^{30}$$
$$+74x^{25}+51x^{20}+30x^{15}+14x^{10}+5x^5+1 \quad (5.10)$$

$$(1+x^5+x^{10}+x^{15}+x^{20}+x^{25})(1+x^{10}+x^{20}+x^{30}+x^{40})(1+x^{25}+x^{50}+x^{75}) =$$
$$x^{140}+x^{135}+2x^{130}+2x^{125}+3x^{120}+4x^{115}+4x^{110}$$
$$+5x^{105}+5x^{100}+6x^{95}+6x^{90}+6x^{85}+6x^{80}+6x^{75}$$
$$+6x^{70}+6x^{65}+6x^{60}+6x^{55}+6x^{50}+6x^{45}+5x^{40}$$
$$+5x^{35}+4x^{30}+4x^{25}+3x^{20}+2x^{15}+2x^{10}+x^5+1 \quad (5.11)$$

We will use these polynomials immediately in Example 5.12. Note that we do not evaluate these polynomials for any particular values of x. A polynomial is a special sort of function that is often used for approximating more complicated functions, but the following example makes a different use of polynomials. A similar method was used on page 65 in expanding $(1+x)^n$ to obtain the binomial coefficients $\binom{n}{r}$.

Example 5.12. With five nickels ($\$.05$), four dimes ($\$.10$), and three quarters ($\$.25$), how many ways is it possible to pay for an item that costs $\$.75$?

1. Solve the problem considering all coins distinct—even those of like value. (For example, it makes a difference not only that we select three nickels but also *which* three we select.)

2. Solve the problem considering coins distinct only if they are of different value. (For example, it makes no difference which three nickels we select.)

Solution of 1. We use equation (5.10). It is easier to see if we write the product on the left side of equation (5.10) in full.

$$(1+x^5)(1+x^5)(1+x^5)(1+x^5)(1+x^5)$$
$$(1+x^{10})(1+x^{10})(1+x^{10})(1+x^{10})$$
$$(1+x^{25})(1+x^{25})(1+x^{25}) \quad (5.12)$$

Since we must pay $\$.75$, we examine the coefficient of x^{75} on the right side of equation (5.10). This coefficient, which is equal to 304, counts the number of ways that it is possible to select either 1 or a power of x from each of the 12 factors of formula (5.12) in such a way that the sum of the chosen exponents of x is 75. This is essentially the same as selecting from the given set of coins a subset of value \$.75, and we see that this can be done in 304 different ways counting coins distinct even if they have the same value.

Solution of 2. We use equation (5.11). The coefficient 6 of x^{75} on the right side of equation (5.11) counts the ways of choosing one term from each of the three factors on the left side of equation (5.11) in such a way that the sum of the exponents of the chosen terms is 75. This computation differs from the previous one using equation (5.10) because in the present case choosing, for example, x^{15} from the first factor on the left of equation (5.11) corresponds *uniquely* to selecting three nickels as part of the \$.75 due; whereas in the previous computation, there are $\binom{5}{3} = 10$ ways of choosing three terms x^5 from the five factors $1 + x^5$ in (5.12), that is, of choosing three of the five nickels. Since the coefficient of x^{75} on the right side of equation (5.11) is 6, that is the number of different ways of selecting \$.75 from the given set of coins considering coins as distinct only if they have different value.

Similarly, the polynomials (5.10) and (5.11) can be used to find the number of ways of using the specified coins to pay any particular amount in place of \$.75.

We have given some examples of the use of polynomials for counting things. This technique plays a major role in the branch of mathematics called *combinatorics* that is concerned with counting. In this chapter, we have taken a few steps in this field. Combinatorics has applications in many areas of science and technology—for example, chemistry, computer science, cryptography, economics, operations research, and statistics.

In the next chapter, we examine a mathematical puzzle that enjoyed worldwide fame in the 1870s. The analysis of this problem introduces the theory of graphs and the algebra of permutations.

Chapter 6

Order and Reorder

The difficult we do immediately; the impossible takes a little longer.
—ANONYMOUS AMERICAN WWII SOLDIER

In the workplace, the demanding boss scolds the slacker who claims that his task is impossible and tells him to try harder. However, in mathematics there are insoluble problems for which trying harder is inappropriate. The ancient problems of *trisecting the angle*, *doubling the cube*, and *squaring the circle* continue to attract would-be solvers despite mathematical proofs that these problems are insoluble.[1] We will see a mathematical proof of the insolubility of another problem whose solution became a worldwide obsession in the 1870s; but first we look at a puzzle discussed 100 years earlier that began a branch of mathematics known as *graph theory*.

The Seven Bridges of Königsberg

The great Swiss mathematician Leonhard Euler (1707–83) discussed and generalized the following problem.

Problem 6.1. Is there a route through the city of Königsberg that crosses each of the seven bridges shown in Figure 6.1(a) exactly once?[2] There are two versions of this question.

1. The route finishes where it starts.

2. The start and finish are in different regions.

Euler abstracts the problem by representing the system of bridges and land regions as shown in Figure 6.1(b). This representation of the problem makes it clear that it is impossible to traverse each bridge exactly once—for the following reasons.

1. On the one hand, if the route finishes where it starts, then, for each region, the route must cross a bridge *toward* that region exactly as many times as it crosses a bridge *away from* that region. Since every bridge is crossed exactly

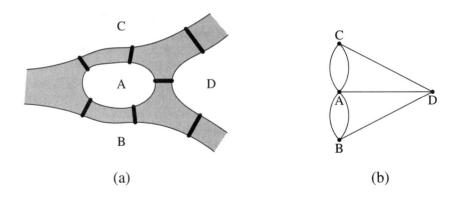

Figure 6.1. The problem of the seven bridges of Königsberg. (a) The city of Königsberg is located partly on the island Kneiphoff (A) at a point where two tributaries of the river Pregel meet; the remaining regions of the city are at B, C, and D. In Euler's time, these four regions were connected, as shown, by seven bridges. Euler discusses the question: Is it possible to walk through the city crossing each bridge exactly once? (b) Euler analyzes the problem using the above graphical representation.

once, it follows that each region is connected to an *even* number of bridges, contrary to the fact that each region is connected to an *odd* number of bridges.

2. On the other hand, if the start and finish are in different regions, then the number of bridges connected, respectively, to the start and finish regions must be odd, and any other region must be connected to an even number of bridges. This is not true because each of the four regions is connected to an odd number of bridges.

A nonempty system of vertexes (points) joined by edges, as in Figure 6.1(b), is called a *graph*. Each edge connects a pair of vertexes. A graph can be defined without the aid of a picture by simply listing which vertexes are incident with which edges.

> A vertex and an edge of a graph are said to be *incident* if the vertex is one of the two endpoints of the edge.

Euler generalizes the problem as follows. *Given a graph, how can we determine whether there exists a route that traverses each edge exactly once?* A route that ends at the same vertex at which it starts is called a *circuit*; a circuit that includes each edge of the graph exactly once is called an *Eulerian circuit*. A route that starts and ends at different vertexes and includes each edge of the graph exactly once is called an *Eulerian path*. The *degree* of a vertex of a graph is defined as the number of edges that meet that vertex. For example, in Figure 6.1(b), the vertex A has degree 5, and the vertexes B, C, and D each have degree 3.

Euler answered his question completely as follows.

Fact 6.2. *A graph has an Eulerian circuit if and only if every vertex has even degree.*

Fact 6.3. *A graph has an Eulerian path if and only if it contains two odd vertexes and all other vertexes are even.*

The two odd vertexes are the beginning and end of the path.

Facts 6.2 and 6.3 confirm that for the seven bridges problem (Figure 6.1), an Eulerian circuit and Eulerian path fail to exist. In other words, no route, whether it returns to its starting point or not, can traverse the seven bridges exactly once.

Hamilton's Problem

It appears that Euler's question has a simple answer. However, a variation of his question, introduced by the Irish mathematician William Rowan Hamilton (1805–65), remains unanswered to this day. Hamilton devised the following puzzle, which is not difficult to solve; the unanswered question concerns a generalization of this puzzle.

Problem 6.4. Following the edges of a regular dodecahedron, find a circuit that contains each vertex exactly once.

> A regular dodecahedron is a solid figure bounded by twelve regular pentagons.

Figure 6.2 is a two-dimensional map of a regular dodecahedron in which a circuit satisfying Hamilton's condition is shown.

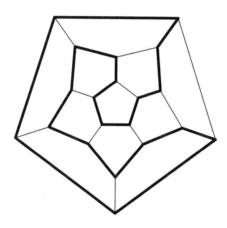

Figure 6.2. Hamilton's problem. The thick lines depict a Hamiltonian circuit.

Given an arbitrary graph, a circuit that begins and ends at the same vertex and meets every vertex of the graph exactly once is called a *Hamiltonian circuit*. For an arbitrary graph, it makes sense to ask whether a Hamiltonian circuit exists. The unanswered question is to find an assertion—like Fact 6.2—that characterizes the graphs for which Hamiltonian circuits exist. In view of the effort mathematicians have given to this question over many years, it is extremely unlikely that there is a characterization of Hamiltonian circuits that is as simple as the characterization of Eulerian circuits given by Fact 6.2.

In Figures 6.3(a) and (c) we see Hamiltonian circuits for graphs containing 16 and 14 vertexes, respectively.

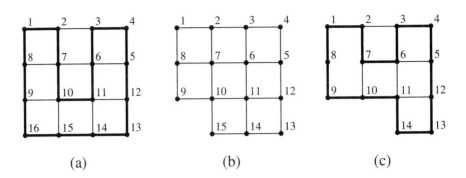

Figure 6.3. Hamiltonian circuits are shown for (a) and (c), graphs with 16 and 14 vertexes, respectively. Does a Hamiltonian circuit exist for the 15-vertex graph (b)?

Problem 6.5. Find a Hamiltonian circuit for the 15-vertex graph shown in Figure 6.3(b).

Solution. There is no Hamiltonian circuit for Figure 6.3(b). In fact, notice that this graph contains more odd-numbered vertexes than even. The vertexes are numbered in such a way that each edge joins an even- to an odd-numbered vertex. Therefore, every circuit—every closed path—must alternate even- and odd-numbered vertexes, and, therefore, must contain an even number of vertexes. It follows that no circuit can contain all 15 vertexes of the graph.

Loyd's Fifteen Puzzle

Sam Loyd (1841–1911) was an American author of puzzles and chess problems. In the late 1870s, one of his creations, the *Fifteen Puzzle*, became a worldwide obsession. At that time in Germany, during meetings of the legislative assembly, the *Reichstag*, some of those distinguished delegates became too distracted by this puzzle to attend to matters of state; in France, shopkeepers threatened their employees with dismissal if they continued to pursue this craze during working hours. Loyd promoted this frenzy by offering $1,000 for the first solution to the puzzle. As we will see, his money was quite safe because a solution is impossible. This impossibility worked against Loyd when he attempted to obtain a U.S. patent for his mechanical puzzle device. The Patent Office declared that if there was no solution there could be no working model to illustrate this impossible device and, therefore, no patent!

But, indeed, the device does exist and can still be found for sale where games and toys are sold. The Fifteen Puzzle consists of a tray of fifteen movable square tiles inscribed with the numbers 1 through 15 as shown in Figure 6.4(a). The tray also contains one empty square. Squares adjacent to the empty square can interchange with it by sliding up, down, right, or left.

Problem 6.6 (Loyd's $1,000 prize problem). Find a sequence of sliding moves that transforms Figure 6.4(a) into Figure 6.4(b).

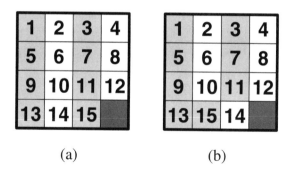

Figure 6.4. Loyd's Fifteen Puzzle. Slide the numbered tiles to transform (a) into (b).

We will show that Loyd's prize problem is impossible, that is, that the arrangement in Figure 6.4(a) cannot be transformed into Figure 6.4(b). We will also show how to determine whether any two given arrangements can be transformed one into the other.

Numbering the tile positions

On the puzzle device there are sixteen positions. One position is blank and the remaining fifteen are occupied tiles bearing the numbers 1 through 15. Independent of the numbers that appear on the tiles, we use the numbers from 1 to 16 to designate the fixed positions that the tiles may occupy. The method of numbering these positions is somewhat arbitrary; we will use the numbering pattern shown in Figure 6.5—the order also used in Figure 6.3(a).

$$
\begin{array}{cccc}
1 & 2 & 3 & 4 \\
8 & 7 & 6 & 5 \\
9 & 10 & 11 & 12 \\
16 & 15 & 14 & 13
\end{array}
$$

Figure 6.5. The order of the tile positions for the puzzle device.

In Problem 6.5, the method of numbering in Figure 6.5 proved more advantageous than numbering each row in increasing order from left to right, and this numbering pattern will also be advantageous in our discussion of the Fifteen Puzzle. Before dealing with Loyd's prize problem, we consider an easier warm-up problem.

Problem 6.7. Regardless of the initial arrangement of the Fifteen Puzzle, show that there exists a sequence of 256 moves of the fifteen numbered tiles whereby every one of the tiles meets every one of the sixteen positions exactly once.

Solution. We interpret the sixteen vertexes of the graph in Figure 6.3(a) as the sixteen positions of the Fifteen Puzzle. Regardless of the initial placement of the blank space, there is a sequence of sixteen moves that causes the blank space to travel around the Hamiltonian circuit in Figure 6.3(a) in the counterclockwise direction. After one such circuit, all the numbered tiles have traveled one step clockwise along the same Hamiltonian circuit. Sixteen counterclockwise revolutions ($16 \times 16 = 256$ moves) of the empty space carries each tile clockwise through the entire Hamiltonian circuit, as shown in Figure 6.6(a). In the course of this movement each tile encounters each position exactly once.

Using the Hamiltonian circuit Figure 6.3(c) in a similar fashion, one can solve part 1 of the following problem, as shown in Figure 6.6(b).

Problem 6.8.

1. Suppose that initially two tiles are located at positions 15 and 16, respectively— as shown in Figure 6.5. Show that there exists a sequence of $14 \times 14 = 196$ moves that leaves the two tiles fixed and such that the thirteen remaining tiles meet every one of the positions 1–14 exactly once.

2. Solve the same problem if the fixed tiles are located at positions 9 and 16, respectively.

Solution. The circuits for parts 1 and 2 are shown in Figures 6.6(b) and (c), respectively.

We conclude this section with a problem that will be useful later.

Problem 6.9. Let x, y, and z be three distinct integers between 1 and 15, inclusive. Show that there exists a sequence of moves that transforms an arbitrary initial arrangement of the Fifteen Puzzle so that the empty square and the tiles numbered x, y, and z are in the positions shown in Figure 6.7, that is, positions 9, 16, and 15.

Solution. Note that our concern here is to show merely that there *exists* such a sequence of moves; we need not find the shortest such sequence. The following proce-

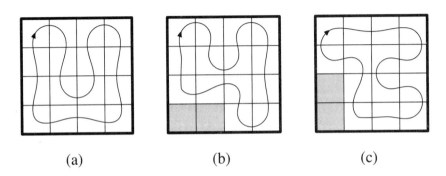

(a)   (b)   (c)

Figure 6.6. Circuits of the Fifteen Puzzle.

dure uses more steps than might be needed in particular cases, but it has the advantage that it is effective for *any* initial arrangement of the tiles.

1. Cycle the tiles according to the circuit in Figure 6.6(a) until the tile marked y is in the lower left corner with the empty space immediately above y.

2. If the tile immediately to the right of y now happens to be x, use one step of the circuit in Figure 6.6(c) to replace this tile with another tile. Note that the tile y remains at the lower left corner.

3. Cycle according to Figure 6.6(b) until the tile x is immediately above tile y, which remains at the lower left corner, as in Figure 6.7.

4. Cycle according to Figure 6.6(c) until the tile z is immediately to the right of tile y. Now x, y, and z are located as shown in Figure 6.7.

5. Move the empty space counterclockwise along the circuit of Figure 6.6(c) until it is located as shown in Figure 6.7. Note that tiles x, y, and z do not move during this step.

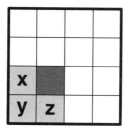

Figure 6.7.

Lists

The *natural* order of the integers 1–15 denotes the normal increasing order. We describe each *arrangement* of the puzzle by listing the numbered tiles in a linear order. Such a *list* contains each integer 1–15 exactly once, but, in general, not in the natural order.

There is more than one way to use a one-dimensional list to represent a two-dimensional arrangement of the numbered tiles. For example, we could list numbers in order from left to right, starting with the top row and proceeding similarly with the second, third, and fourth rows. This representation seems natural and, indeed, some authors have used it. However, the subsequent explanation is somewhat simpler if we use the pattern shown in Figure 6.8(a) in which we order the first row from left to right, the second from right to left, and so on, reversing the direction of each subsequent row. We will say that the list obtained in this way from an arrangement of the Fifteen Puzzle is *associated* with that arrangement.

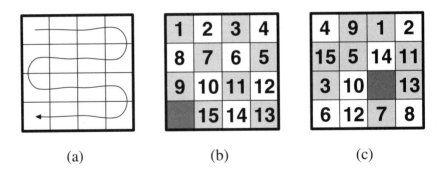

(a)   (b)   (c)

Figure 6.8. For each arrangement of the puzzle, the associated list is determined by listing the numbered tiles in the order shown in (a). The list associated with the arrangement (b) is the natural order of the integers 1–15, and the list associated with the arrangement (c) is 4, 9, 1, 2, 11, 14, 5, 15, 3, 10, 13, 8, 7, 12, 6.

This method of using a list to represent an arrangement of the tiles is related to our method, described in the previous section, of numbering the fixed tile positions. If the nth number in a list is m, it means that the tile bearing the number m is located either at the position n or at position $n + 1$. The reason for the ambiguity is that the list ignores the position of the empty square. However, we will see that, despite this ambiguity, the knowledge of the associated lists of two arrangements is sufficient for determining whether it is possible to transform the one arrangement into the other. In the following discussion, when we see exactly how manipulating the puzzle affects the associated list, we can use the lists as proxies for the arrangements.

The list associated with Figure 6.4(a) is

$$1, 2, 3, 4, 8, 7, 6, 5, 9, 10, 11, 12, 15, 14, 13 \qquad (6.1)$$

and for Figure 6.4(b) it is the same except that 15 and 14 are interchanged. The list associated with the arrangement of Figure 6.8(c) is

$$4, 9, 1, 2, 11, 14, 5, 15, 3, 10, 13, 8, 7, 12, 6 \qquad (6.2)$$

Inversions

Each pair of numbers of a list that are not in the natural order—that is, each pair for which the larger number comes in the list before the smaller number—is called an *inversion*. To count as an inversion, a pair of numbers need not be adjacent in the list. For example, in the list (6.2) the pair 9, 3 counts as an inversion because 9 is listed before 3, contrary to the normal order. For the list (6.1), associated with Figure 6.4(a), there are exactly nine inversions:

$$(8,7), (8,6), (8,5), (7,6), (7,5), (6,5), (15,14), (15,13), (14,13)$$

The list associated with Figure 6.4(b) has eight inversions: all of the above except $(15, 14)$.

Figure 6.9 shows the computation of the total inversions for the list (6.2). This chart contains a triangular array of 105 entries, one for each pair of distinct integers between 1 and 15. Each entry is 0 or 1 depending on whether the corresponding pair occurs in the list (6.2) in the normal ascending order. Each entry that is equal to 1 indicates an inversion in the list (6.2). Thus, the total, 43, of all entries in the triangular array is equal to the total number of inversions in the list (6.2).

	1	2	3	4	5	6	7	8	9	10	11	12	13	14
15	0	0	1	0	0	1	1	1	0	1	0	1	1	0
14	0	0	1	0	1	1	1	1	0	1	0	1	1	
13	0	0	0	0	0	1	1	1	0	0	0	1		
12	0	0	0	0	0	1	0	0	0	0	0			
11	0	0	1	0	1	1	1	1	0	1				
10	0	0	0	0	0	1	1	1	0					
9	1	1	1	0	1	1	1	1						
8	0	0	0	0	0	1	1							
7	0	0	0	0	0	1								
6	0	0	0	0	0									
5	0	0	1	0										
4	1	1	1											
3	0	0												
2	0													
Total inversions	2	2	6	0	3	9	7	6	0	3	0	3	2	0
Grand total	43													

Figure 6.9. Inversions of the list 4, 9, 1, 2, 11, 14, 5, 15, 3, 10, 13, 8, 7, 12, 6. This computation shows that 43 is the number of inversions. Hence this list—which corresponds to Figure 6.8(c)—is odd.

Definition 6.10. A list is said to be *odd* or *even* depending on whether the total number of inversions of the list is odd or even.

This is more than just an *ad hoc* definition. Even and odd lists are important elsewhere in mathematics.

Definition 6.11. An arrangement of the Fifteen Puzzle is said to be odd or even depending on whether the associated list is odd or even.

The above definition depends on our method of associating lists with arrangements of the Fifteen Puzzle. A different method of associating lists with arrangements would lead to a different definition of even and odd arrangements.

We can now state the principal result concerning the Fifteen Puzzle. The *parity* of an arrangement is its even-or-odd state.

Proposition 6.12. *One arrangement of the Fifteen Puzzle can be transformed into another arrangement if and only if both arrangements have the same parity.*

There are two halves of this assertion.

1. Two arrangements with the same parity *can* be transformed into each other. This turns out to be the more difficult half of the assertion.

2. Two arrangements with different parities *cannot* be transformed into each other. This half of the assertion implies that the solution for which Loyd offered prize money is impossible because Loyd's problem requires transforming an arrangement of odd parity into one of even parity.

The following example illustrates the fact that a single move does not change the parity of an arrangement.

Example 6.13. In Figure 6.8(c), let us examine the effect on the associated list produced by moving the square marked 14 downward, thereby trading places with the empty square.

The number 14 leapfrogs over the four numbers, 5, 15, 3, and 10, transforming the list (6.2) into

$$4, 9, 1, 2, 11, 5, 15, 3, 10, \mathbf{14}, 13, 8, 7, 12, 6$$

The number of inversions is altered as follows.

	Pair	Inversions
Before the move	14, 5	1
	14, 15	0
	14, 3	1
	14, 10	1
After the move	5, 14	0
	15, 14	1
	3, 14	0
	10, 14	0

Thus, in this example, the total number of inversions decreases by two, from 3 to 1. The inversion state of every other pair is unaffected. Each "leap" either increases or decreases the total number (four) of inversions by one, thus changing the parity of the total. But since, in this example, there are an even number of leaps, the parity of the total is unaffected.

Example 6.13 illustrates the general fact stated by the following proposition.

Proposition 6.14. *If an arrangement of the Fifteen Puzzle is altered by interchanging the empty square with an adjacent numbered square, then the parity of the arrangement does not change.*

Proof. Moving a square to the left or right does not change the associated list, and hence the parity of the arrangement remains unchanged. On the other hand, moving a square up or down does change the associated list, but, nevertheless, we will see that the parity of the arrangement is again unchanged.

Consider the process one step at a time. As a number leapfrogs over *one* adjacent number in a list, the inversion state of the transposed pair, the "leaper" and the "leapee," is reversed while the inversion states of all other pairs are unaffected by this transposition. Therefore, the parity of the total number of inversions reverses (odd to even, or even to odd). A second leap—indeed, any even number of leaps—brings the parity back to its original state.

As a second example, consider the effect on the parity of the list in Figure 6.8(c) if we move the 7 upwards one square. In the associated list, 7 leaps over 8 and 13. Again, since there is an even number of leaps, the parity of the total inversions is unchanged.

In fact, for any arrangement, a single move upward or downward entails a leap over an *even* number of numbers in the associated list. Therefore, the parity of the arrangement is unaffected by any single move and, indeed, any sequence of moves.

□

It follows that the problem posed by Loyd is impossible because it requires us to transform an odd arrangement into an even one.

What arrangements are possible?

We have proved only the second half of Proposition 6.12. We have shown that two arrangements cannot be transformed one into the other if they have different parities. The remaining task is to show that two arrangements can be so transformed if they have the same parity. One can become convinced of this fact by means of a practical demonstration. To participate fully in this demonstration, one needs to work with a version of the puzzle that permits removal of the tiles from the tray so that we can form an arbitrary initial position. This can be done by writing the numbers 1–15 on slips of paper and placing them in a square array—for example, on a 4×4 portion of a chessboard. After a number of trials, one becomes convinced of the following fact.

Fact 6.15. *An arbitrary given arrangement can be transformed either into the odd arrangement of Figure 6.4(a) or into the even arrangement of Figure 6.4(b).*

For example, it is possible to transform the odd arrangement of Figure 6.8(c) into Figure 6.4(a), which is also odd.

It follows from Fact 6.15 that any two arrangements of the same parity can be transformed one into the other. In fact, suppose that A and B are both odd arrangements. From Fact 6.15, both A and B can be transformed into Figure 6.4(a). To transform A into B, first transform A into Figure 6.4(a); then reverse the moves that transform B into Figure 6.4(a). A similar argument works if both A and B are even.

But how can we be completely certain that Fact 6.15 is true? A convincing argument must show exactly how to transform an arbitrary arrangement either into Figure 6.4(a) or Figure 6.4(b). This task seems daunting because so many initial arrangements are possible. Therefore, we will follow a different path in proving that two arrangements can be transformed, one into the other, if they have the same parity. Our goal is to show that for any two arrangements of the same parity *there exists* a sequence of legal moves that transforms the one arrangement into the other. There are two difficulties that finding such a transformation might entail.

1. The transformation might prescribe a very large number of moves. This would make it difficult to carry out the transformation using the physical puzzle device.

2. The transformation might be difficult to understand. It might require many pages of text for its description, and it might be difficult to see that it works for all possible initial arrangements.

The transformation that we find in the proof of Proposition 6.26 is described in a few lines, but it requires too many moves to be of practical use; that is, the proof overcomes difficulty 2 but not difficulty 1. This is satisfactory because we first want to know that a transformation exists before we are concerned with the most efficient manner of achieving the transformation.

To continue this proof, we need to introduce the concept of *permutation*.

Permutations

Roughly speaking, *permutations* are methods of altering the order of the elements of a list. A permutation *maps* a *source* list into an *image* list.[3]

We introduce the concept of permutation by means of an example.

Example 6.16 (the mixer dinner). Suppose that a hostess is planning a dinner party for 15 people. The dinner will be served in five courses: appetizer, soup, salad, entrée, and dessert. To encourage more interaction between the guests, the hostess plans that the guests should move to new seats at the beginning of each course. She can organize this in two different ways.

1. She can prepare five different lists showing the seat each guest should occupy for each of the five courses.

2. She can prepare one list showing the initial seating for the appetizer course. Then she can prepare four charts, corresponding to the four reseatings between the courses. These charts show for each guest the name of the guest whose seat he should occupy for the next course. For example, one of the charts might show, in part, that Ada should move for the salad course to the seat that was previously occupied by Ben during the soup course. Each of the four charts describes a *permutation* of the guests. If we assign the numbers 1–15 in any convenient manner to the guests, then the following definition applies.

Definition 6.17. A *permutation* of the numbers 1–15 is a mapping that associates each of these numbers, one-for-one, with the same numbers possibly in a different order.

We can generalize the above definition by replacing 15 with an arbitrary natural number.

The following is an example of the notation for a permutation.

$$\begin{pmatrix} 1 & 2 & 3 & 4 & 5 & 6 & 7 & 8 & 9 & 10 & 11 & 12 & 13 & 14 & 15 \\ 4 & 9 & 1 & 2 & 11 & 14 & 5 & 15 & 3 & 10 & 13 & 8 & 7 & 12 & 6 \end{pmatrix} \qquad (6.3)$$

To define a particular permutation, it is sufficient to show how it maps one particular source list because this shows the image of each of the numbers 1–15, also called *elements*, under the permutation. The top row represents a source list and the bottom row is an image list with the corresponding numbers one under the other. In terms of the dinner party, the permutation (6.3) says that guest number 1 moves to the seat vacated by guest number 4, and so on. Note that guest number 10 keeps his previous seat.

The hostess may find the permutation method especially attractive if she chooses permutations that can be stated simply. For example, she might use the rule that at the end of each course the ladies keep their seats, and each gentleman moves to the seat of the gentleman to his right. Thereby, the four between-courses permutations are identical, and no charts or complicated instructions to the guests are necessary.

We have seen that the list associated with an arrangement of the Fifteen Puzzle undergoes changes when a tile is moved. We can specify these changes by means of a permutation of the integers 1–15. In particular, in Example 6.13 the permutation is the following:

$$\begin{pmatrix} 4 & 9 & 1 & 2 & 11 & 14 & 5 & 15 & 3 & 10 & 13 & 8 & 7 & 12 & 6 \\ 4 & 9 & 1 & 2 & 11 & 5 & 15 & 3 & 10 & 14 & 13 & 8 & 7 & 12 & 6 \end{pmatrix} \qquad (6.4)$$

This permutation can also be written as follows:

$$\begin{pmatrix} 1 & 2 & 3 & 4 & 5 & 6 & 7 & 8 & 9 & 10 & 11 & 12 & 13 & 14 & 15 \\ 1 & 2 & 10 & 4 & 15 & 6 & 7 & 8 & 9 & 14 & 11 & 12 & 13 & 5 & 3 \end{pmatrix} \qquad (6.5)$$

Inverse permutation. Corresponding to each permutation, there is an *inverse* permutation, the permutation that restores the original order. The permutation that is inverse to (6.3) is

$$\begin{pmatrix} 4 & 9 & 1 & 2 & 11 & 14 & 5 & 15 & 3 & 10 & 13 & 8 & 7 & 12 & 6 \\ 1 & 2 & 3 & 4 & 5 & 6 & 7 & 8 & 9 & 10 & 11 & 12 & 13 & 14 & 15 \end{pmatrix}$$

For every permutation, there is a corresponding inverse permutation.

Definition 6.18 (transposition). A permutation that interchanges two distinct numbers and leaves the other numbers unchanged is called a *transposition*.

In other words, a transposition *swaps* two specified elements. The notation shown in (6.3) is cumbersome for representing a transposition. The permutation that interchanges a and b is represented $(a\ b)$ or $(b\ a)$. For example, $(1\ 2)$ represents the permutation that interchanges 1 and 2.

Definition 6.19 (cyclic permutation). A permutation that replaces b with a, c with $b, \ldots, z$ with y, and a with z is called a *cyclic* permutation, or simply a *cycle*, and is written $(a\ b \cdots y\ z)$.[4]

Visualize $a, b, c, \ldots, z$ seated at a round table in clockwise alphabetical order with z seated next to a. The permutation in Definition 6.19 causes each element to move one place in the clockwise direction.

Sometimes we wish to specify the number of elements in a cycle. For example, the permutation $(1\ 2\ 3)$ is the 3-cycle that replaces 2 with 1, 3 with 2, and 1 with 3. The notation for cyclic permutations is consistent with the notation used for transpositions because a transposition, for example $(1\ 2)$, is a special kind of cyclic permutation, a 2-cycle. The permutation (6.4) (which is the same as the permutation (6.5)) is cyclic and can be written $(14\ 5\ 15\ 3\ 10)$. This cyclic permutation can also be written $(5\ 15\ 3\ 10\ 14)$ or $(15\ 3\ 10\ 14\ 5)$, and so on. In this notation, elements not listed are unchanged.

The algebra of permutations

If $\mathcal{M}$ is the the list obtained by mapping a list $\mathcal{L}$ by means of a permutation p, then we write $\mathcal{M} = p\mathcal{L}$. We can now apply a permutation q to the list $\mathcal{M}$, obtaining a list $\mathcal{N}$ satisfying $\mathcal{N} = q\mathcal{M} = q(p\mathcal{L})$. Without ambiguity, we can omit the parentheses from the right side of this equation and write $qp\mathcal{L}$. The permutation that expresses the relation between $\mathcal{N}$ and $\mathcal{L}$ is denoted as a product: qp. Note that when the product permutation qp is applied to a list $\mathcal{L}$, first the right factor p is applied to $\mathcal{L}$, and then the permutation q is applied to the result.

The result of applying first p and then q is different, in general, from applying first q and then p. In this respect, the multiplication of permutations differs from the ordinary multiplication of numbers. For example, the result of applying first the transposition $(1\ 2)$ and then $(1\ 3)$ is equal to the cyclic permutation $(1\ 2\ 3)$, but the result of applying first $(1\ 3)$ and then $(1\ 2)$ is the 3-cycle $(1\ 3\ 2)$. Figure 6.10—we will call it a *trace* diagram—verifies these facts. In Figure 6.10, we trace the elements remembering that the permutations are applied *from right to left*.

Product	Trace			3-cycle
$(1\ 3)(1\ 2)$	$1 \rightarrow 2$	$2 \rightarrow 1 \rightarrow 3$	$3 \rightarrow 1$	$(1\ 2\ 3)$
$(1\ 2)(1\ 3)$	$1 \rightarrow 3$	$3 \rightarrow 1 \rightarrow 2$	$2 \rightarrow 1$	$(1\ 3\ 2)$

Figure 6.10. Trace diagram that shows that $(1\ 2)(1\ 3)$ is different from $(1\ 3)(1\ 2)$.

Thus, putting $p = (1\ 2)$ and $q = (1\ 3)$ and using Figure 6.10, we may write

$$qp = (1\ 3)(1\ 2) = (1\ 2\ 3) \qquad pq = (1\ 2)(1\ 3) = (1\ 3\ 2)$$

The permutation that does nothing—that leaves every list unchanged—is called the identity permutation and is written e. The inverse of a permutation p is denoted p^{-1}. A permutation p is the inverse of q if $pq = e$.

Every transposition t is its own inverse because a second application of t reverses the interchange of the first application. The inverse of a cyclic permutation $(a\ b\ c \ldots x\ y\ z)$ is the cyclic permutation of the same elements in the reverse order: $(z\ y\ x \ldots c\ b\ a)$. For example, by tracing elements, one can see that the inverse of $(1\ 2\ 3)$ is $(1\ 3\ 2)$.

The inverse of a product $(pq)^{-1}$ is equal to the product of the inverses in reverse order, $q^{-1}p^{-1}$, as can be seen from the following calculation:

$$(q^{-1}p^{-1})(pq) = q^{-1}(p^{-1}p)q = q^{-1}eq = q^{-1}q = e$$

Even and odd permutations

Fact 6.20. *Every permutation can be written as a product of cyclic permutations with no element in common.*

For example, the permutation (6.3) can be written as follows:[5]

$$(1\ 4\ 2\ 9\ 3)(5\ 11\ 13\ 7)(6\ 14\ 12\ 8\ 15) \tag{6.6}$$

In general, to find such a product, trace a chain, as in Figure 6.10, starting with an arbitrary element—for example, starting with 1. Eventually, the chain returns to its starting element, thus defining the first cyclic permutation. For example, for the permutation (6.3) we have

$$1 \rightarrow 4 \rightarrow 2 \rightarrow 9 \rightarrow 3 \rightarrow 1$$

Continuing, we find an element not included in the above cycle, and similarly construct a cycle starting with that element, and so on. If no starting element can be found that is not included in one of the cycles previously found, then the process is finished and we have expressed the given permutation as a product of cycles.

Fact 6.21. *Every permutation can be represented as product of transpositions.*

For example, the permutation (6.3) is equal to

$$(6\ 15)(6\ 8)(6\ 12)(6\ 14)(5\ 7)(5\ 13)(5\ 11)(1\ 3)(1\ 9)(1\ 2)(1\ 4)$$

To verify this example, use the fact that permutation (6.3) is equal to the product of cycles (6.6); trace through each transposition from right to left using the technique shown in Figure 6.10.

Proposition 6.22. *A transposition changes the parity of an arbitrary list from odd to even or from even to odd.*

Proof. We must show that a transposition changes the number of inversions of the list from odd to even or even to odd. Let m and n be the two distinct numbers that are interchanged; in other words, the transposition is $(m\ n)$. Let L be an arbitrary list. There are a certain number $k \geq 0$ of integers that lie between m and n in the list L. The transposition $(m\ n)$ can be achieved by a succession of leaps as in the proof of Proposition 6.14.

1. The integer m leaps over each of the k intervening integers between m and n. Each leap changes the parity of the list because it increases or decreases the total number of inversions by 1.

2. The integers m and n are now adjacent. Interchange them, thereby changing the parity once again.

3. The integer n leaps over the k intervening integers—the same as traversed by m in step 1 but in the opposite direction.

Since there are as many parity changes in step 3 as in step 1, the total number of parity changes is odd, confirming the assertion of the proposition.   □

From Fact 6.21 and Proposition 6.22, we see that the totality of permutations can be divided into two non-overlapping categories:

Even permutations are expressible as a product of an even number of transpositions. Application of an even permutation preserves the parity of an arbitrary list.

Odd permutations are expressible as a product of an odd number of transpositions. Application of an odd permutation reverses the parity of an arbitrary list.

Fact 6.23. *A cyclic permutation is odd or even depending on whether the cycle contains an even or odd number of elements.*

For example, the cycle (2 3 7) is an even permutation because, using a trace diagram like Figure 6.10, we can verify $(2\ 3\ 7) = (2\ 7)(2\ 3)$. Moreover, the cycle (2 3 7 8) is odd because $(2\ 3\ 7\ 8) = (2\ 8)(2\ 7)(2\ 3)$.

Fact 6.24. *Every product of two transpositions can be expressed as a 3-cycle or a product of two 3-cycles.*

In fact, $(a\ b)(c\ d) = (a\ c\ d)(a\ b\ d)$ and $(a\ b)(a\ c) = (a\ c\ b)$. If the two transpositions are identical, then their product is the identity permutation, the permutation that changes nothing.

Fact 6.25. *Every even permutation can be expressed as a three-element cycle or a product of three-element cycles.*

In fact, every even permutation can be expressed as a product of an even number of transpositions and each pair of transpositions, according to Fact 6.24 can be expressed as a 3-cycle or a product of 3-cycles.

We know that legal moves of the Fifteen Puzzle correspond to even permutations.

Proposition 6.26. *For the Fifteen Puzzle, every even permutation of the list corresponding to the initial position can be achieved by a suitable sequence of moves.*

Proof. Since, according to Fact 6.25, every even permutation can be expressed as a product of 3-cycles, it is sufficient to show that, given an arbitrary initial arrangement, every 3-cycle can be realized by a suitable sequence of legal moves of the Fifteen Puzzle.

1. Let $(x\ y\ z)$ be an arbitrary 3-cycle. The solution to Problem 6.9 shows that it is possible to transform an arbitrary initial arrangement so that the empty square and the tiles x, y, and z are located as shown in Figure 6.7.

2. Furthermore, it is easy to transform Figure 6.7 into Figure 6.11 by executing a circuit of the four squares in the lower left quarter of the puzzle.

3. Finally, we transform Figure 6.11 by executing in reverse the sequence of moves described above in step 1.

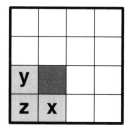

Figure 6.11.

The result of these three steps is the initial arrangement with x replaced by y, y by z, and z by x; all the other tiles are exactly as they were in the initial arrangement. It follows that, by these three steps, we have realized the 3-cycle $(x\ y\ z)$. ☐

In this chapter we have seen the mathematics of Loyd's Fifteen Puzzle. In the next chapter, we begin by looking at another device in which counters are moved in a frame—the abacus, an ancient tool for counting and calculating. Casting out nines, an old method for checking arithmetic calculation, leads to modular arithmetic, a topic that will be used later in Chapter 18 to develop a method of secret communication.

Chapter 7

Outcast

And all arithmetic and calculation have to do with number?
Yes.
And they appear to lead the mind toward truth?
Yes, in a very remarkable manner.

—PLATO, *The Republic*

There is a very old method for checking arithmetic calculations called *casting out nines*. In this chapter we begin by describing this technique. Many find the procedure mysterious and ask the question, "Why does it work?" The answer is the centerpiece of this chapter. It leads to an important extension of ordinary arithmetic called *modular arithmetic*, a concept that underlies cyclic patterns of ordinary life such as the hours of the day and the days of the week.

As a tool of calculation, casting out nines has the shortcoming that it is unable to detect one of the most common arithmetic errors—transposition of digits. However, it is a charming curiosity and, more important, its inner workings lay bare some significant mathematics.

The use of the word "cast" is a clue that the method of casting out nines dates from a distant time when the *abacus* was the preferred tool of calculation. The meaning of *cast* as used here is similar to the meaning *to throw* because the calculation is carried out by moving, that is, *casting*, the beads. (See Figure 7.1.) The abacus is a computing device that consists of a frame holding several parallel rods strung with movable beads—in Latin *calculi* or pebbles. Reckoning with pebbles is the ancient meaning of *calculate*.

Casting Out Nines

We now define the process of casting out nines.

Definition 7.1. To *cast out nines* from a nonnegative integer means to alternate the following two processes until a nonnegative integer less than 9 is obtained. The processes are:

 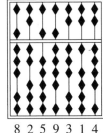

0 0 0 0 0 0 0   8 2 5 9 3 1 4

Figure 7.1. The Chinese abacus, the *suan-pan*, consists of two *decks* separated by a *beam*. Each rod in the upper deck holds two beads, each counted as five; and each rod in the lower deck holds five beads, each counted as one. Beads are considered counted if they are moved toward the beam. The left abacus is set to the number 0; the right abacus is set to 8,259,314. The Japanese abacus, the *soroban*, has only one row of beads in its upper deck.

1. Cross out any digits that add to 9 or a multiple of 9. (If we fail to cross out some digits because we fail to notice that they add to a multiple of 9, the method still succeeds, but may involve slightly more computation.)

2. Add the remaining digits.

Example 7.2. Cast out nines from the number $8,743,654$.

Solution.

- Cross out digits: $8,\cancel{743,65}4 = 874$. The digits crossed out sum to 18 which is a multiple of 9.

- Add the remain digits: $8 + 7 + 4 = 19$.

- Cross out digits: $1\cancel{9} = 1$. No more steps are needed. Casting out nines from $8,743,654$ yields 1.

To **cast out nines from a negative integer**, which we denote $-n$, first apply the above procedure to the positive integer n, and then subtract the result from 9. Referring to Example 7.2, casting out nines from $-8,743,654$ yields $9 - 1 = 8$.

Checking arithmetic calculations

General pattern

An arithmetic calculation consists of certain input numbers and an output (or answer). Casting out nines can be used to check arithmetic calculations involving addition, subtraction, and multiplication. The general pattern is described below. Later, we will discuss the specific procedures for checking addition and multiplication.

1. Compute a test number based on the numerical inputs. This number is obtained by casting out nines by a procedure that depends on the particular calculation.

2. Cast out nines from the output.

3. The numbers obtained in step 1 and step 2 must be identical. If they are not, then there is an error in calculation. [However, even if the numbers are the same, the calculation might still be incorrect. An answer chosen at random has a probability of 1 in 9 (1/9) of passing this test.]

This procedure applies to arithmetic calculations in general. Now we will give specific versions of these three steps, first for addition and then for multiplication.

Checking addition

We will use the following problem to illustrate the method for checking addition. The input numbers to be added are called *summands*.

$$\text{Summands} \begin{cases} 4,362 \\ 2,049 \\ +\quad 6,931 \end{cases} \tag{7.1}$$
$$\text{Sum} \qquad 13,342$$

To check the addition, we proceed as follows.

1. Carry out the following procedure for finding a test number from the inputs— the summands.

   (a) Cross out digits that sum to nine or a multiple of nine even if they belong to *different* summands.

   $$4,\cancel{362}$$
   $$2,04\cancel{9}$$
   $$\cancel{6,93}1$$

   The crossed out digits sum to 36, which is a multiple of 9. From the remaining numbers, we could also have crossed out 4, 4, and 1—their sum is 9—but this would not affect the outcome. The method succeeds even if we fail to cross out some digits that add to a multiple of 9.

   (b) Add the remaining digits belonging to *all* of the summands.

   $$4+2+2+0+4+1 = 13$$

   (c) At this point, since no subset of the digits of the number 13 add to 9 or a multiple of 9, it is only necessary to add the digits.

   $$1+3 = \mathbf{4}$$

2. Cast out nines from the sum.

    (a) Cross out digits that sum to 9 or a multiple of 9.

$$13,\cancel{342} \to 13$$

    (b) Add the digits, obtaining

$$1+3 = \mathbf{4}$$

3. The numbers obtained in step 1 and step 2 must be identical. If they are not, then there is an error in addition. Since we obtained the number 4 in both steps, the addition (7.1) passes the check.

Checking multiplication

We will use the following problem to illustrate the method for checking multiplication. Recall (page 24) that the input numbers to be multiplied are called *factors*.

$$
\begin{array}{r}
632 \\
\times \quad 845 \\
\hline
3160 \\
2528 \\
5056 \\
\hline
534040
\end{array}
\qquad (7.2)
$$

1. Cast out nines to obtain a test number from the factors as follows.

    (a) Cast out nines from the first factor.

$$\cancel{63}2 \to 2$$

    (b) Cast out nines from the second factor.

$$8\cancel{45} \to 8$$

    (c) Multiply the numbers obtained in (a) and (b).

$$2 \times 8 = 16$$

    (d) Cast out nines from the result obtained in (c). Since there are no digits of 16 adding to 9 or a multiple of 9, it is only necessary to add the digits of 16, the number obtained in (c).

$$1+6 = \mathbf{7}$$

2. Cast out nines from the product.

(a) Cross out digits that add to 9 or a multiple of 9.

$$534040 \to 3040$$

(b) Add the remaining digits of the product.

$$3+0+4+0 = 7$$

3. The numbers obtained in step 1 and step 2 must be identical. If they are not, then there is an error in multiplication. Since we obtained the number 7 in both steps, the multiplication (7.2) passes the check.

This concludes our discussion of *how* casting out nines works. Soon we will see *why* it works, but first we must introduce some important concepts that we will use extensively in Chapters 16, 17, and 18. This is a rather lengthy digression, but it is a necessary preparation for the bright idea behind casting out nines.

Modular Arithmetic

The key concept behind casting out nines is called *modular arithmetic*. Roughly speaking, modular arithmetic replaces the integers with their remainders after division by a certain fixed natural number called the *modulus*. For the method of casting out nines, the appropriate modulus is the number 9.

Odd and even

There is a familiar instance of modular arithmetic—the arithmetic of even and odd. In general, the parity of a sum/product is determined by the parities of the summands/factors. For example, *the sum of an even integer and an odd integer is odd.* This and other similar plausible facts—we omit the proofs—validate the tables of addition and multiplication in Figure 7.2. These tables justify calling this system a miniature arithmetic.

Addition		
+	Even	Odd
Even	Even	Odd
Odd	Odd	Even

Multiplication		
×	Even	Odd
Even	Even	Even
Odd	Even	Odd

Figure 7.2. Addition and multiplication tables of even and odd.

Red, blue, and green

Even and odd provide a miniature system of arithmetic with just two "numbers," but there are no English words like even and odd to describe the analogous system with

modular. The mathematical meaning of the word *modular* does not relate to the common usage of *module* as a standardized component of a system. The mathematical usages—there are several—are based on a different meaning of module: a *standard* or a *unit of measure*. *Module* is the English form of the Latin word *modulus*. Mathematics also uses the Latin form *modulo*, meaning *with respect to the modulus*. In the study of complex numbers, *modulus* has a different meaning. See Definition 15.5 on page 217.

arithmetic. The word *arithmetic* is used broadly to denote systems of mathematical objects with operations that resemble ordinary addition and/or multiplication. A well-behaved arithmetic should satisfy most of the following laws for arbitrary elements $a, b,$ and c. Modular arithmetic satisfies all of these laws.

The Commutative Laws.
Addition:

$$a+b = b+a$$

Multiplication:

$$ab = ba$$

The Associative Laws.
Addition:

$$a + (b+c) = (a+b) + c$$

Multiplication:

$$a(bc) = (ab)c$$

The Distributive Law.

$$a(b+c) = ab + ac$$

parity. The even or odd state of an integer is called its *parity*. If two integers are both odd or both even, they are said to have the same parity; if one is odd and one even, they have opposite parity.

Pronunciation.
Congruent.
KONG'·gru·ent.

In geometry, *congruent* has a different meaning. See Definition 12.2 on page 175.

three numbers. To overcome this lack, let us say that a nonnegative integer is red, blue, or green depending on whether the remainder after division by three is 0, 1, or 2, respectively. We extend this definition to the negative integers by replacing 0, 1, and 2 by -3, -2, and -1, respectively.

$$\cdots \quad -4 \quad -3 \quad -2 \quad -1 \quad 0 \quad 1 \quad 2 \quad 3 \quad 4 \quad \cdots$$
$$\cdots \quad \text{G} \quad \text{R} \quad \text{B} \quad \quad \text{G} \quad \text{R} \quad \text{B} \quad \text{G} \quad \text{R} \quad \text{B} \quad \cdots$$

We can define an arithmetic of red, blue, and green that resembles the one of even and odd. For example, if we add a number with remainder 1 (e.g., 7) to another number with remainder 2 (e.g., 11), we obtain a number ($7 + 11 = 18$) that is divisible by 3; in other words, using our definition of red, blue, and green numbers we have

$$B + G = R$$

Extending this example, the complete addition and multiplication tables for R, B, and G are shown in Figure 7.3.

Addition				Multiplication		
+	R	B	G	**×**	R	B	G
R	R	B	G	R	R	R	R
B	B	G	**R**	B	R	B	G
G	G	R	B	G	R	G	B

Figure 7.3. Addition and multiplication tables of R, B, and G. *Example:* The boldface **R** in the addition table corresponds to the sum $B + G = R$.

Congruence

The systems of even and odd and of red, blue, and green inherit familiar properties of the arithmetic of integers—for example, the property that the order in which we add or multiply two numbers does not affect the sum/product. These systems are instances of modular arithmetic. To prepare for the general definition of modular arithmetic, we define what it means for one number to be congruent to another with respect to a modulus.

Definition 7.3. Let a and b be integers, and let m be a natural number. If there exists an integer k such that $a - b = km$, then we say a *is congruent to* b *modulo* m, and we write

$$a \equiv b \pmod{m}$$

Stated less formally, this means that a and b leave the same remainder after division by m.

The symbol $\equiv$ has other meanings in other contexts (page 17), but the mod symbol is a sure indication that congruence is intended. The congruence relation is similar to equality in certain respects that are stated in the following fact.

Fact 7.4. *Let a, b, c, and d be integers, and let m be a natural number. Suppose the following two congruences are true:*

$$a \equiv b \pmod{m} \qquad\qquad c \equiv d \pmod{m}$$

Then the following congruences are also true:

$$a + c \equiv b + d \pmod{m} \tag{7.3}$$
$$a - c \equiv b - d \pmod{m} \tag{7.4}$$
$$ac \equiv bd \pmod{m} \tag{7.5}$$

The proof of this plausible fact will be omitted. Let us verify that it is true for the following example.

Example 7.5. Verify Fact 7.4 with $a = 3$, $b = 21$, $c = 14$, $d = 5$, and $m = 9$.

Solution. To verify the hypotheses of Fact 7.4, we must check the correctness of the congruences

$$3 \equiv 21 \pmod{9} \qquad\qquad 14 \equiv 5 \pmod{9}$$

Indeed, in both cases subtracting the left side from the right gives a multiple of 9.

To verify the conclusions of Fact 7.4 we must show that the following congruences hold.

$$3 + 14 \equiv 21 + 5 \pmod{9}$$
$$3 - 14 \equiv 21 - 5 \pmod{9}$$
$$3 \times 14 \equiv 21 \times 5 \pmod{9}$$

By performing the arithmetic on the right and left sides, we see that it is sufficient to show:

$$17 \equiv 26 \pmod{9}$$
$$-11 \equiv 16 \pmod{9}$$
$$42 \equiv 105 \pmod{9}$$

We complete the verification of Fact 7.4 by noting that, in each of the three cases, subtracting the right side from the left gives a multiple of 9 as required by Definition 7.3.

We will make extensive use of the following consequence of Fact 7.4.

Proposition 7.6. *Let a, b, and m be integers such that $a \equiv b \pmod{m}$. Then for any natural number n, we have*

$$a^n \equiv b^n \pmod{m} \tag{7.6}$$

Proof. For example, putting $c = a$ and $d = b$ in equation (7.5), we see that

$$a^2 \equiv b^2 \quad (\text{mod } m) \tag{7.7}$$

follows from

$$a \equiv b \quad (\text{mod } m)$$

The general result follows from repeated application of equation (7.5). □

One interpretation of Fact 7.4 is that the congruence relation ($\equiv$) resembles equality ($=$) in many respects. For example, equation (7.3) is similar to the assertion that *equals added to equals are equal.*

The following two examples show that equation (7.6) of Proposition 7.6 can be used to reduce a seemingly difficult computation to an easy one.

Example 7.7. Show that $23^{100} \equiv 1 \pmod{11}$.

Solution. Since we have $23 \equiv 1 \pmod{11}$ and $1^{100} = 1$, the result follows from equation (7.6).

Example 7.8. Show that $17^4 \equiv 1 \pmod{12}$.

Solution 1. First compute

$$17^4 = 17 \times 17 \times 17 \times 17 = 83,521$$

After dividing $83,521$ by 12 we obtain a remainder of 1. This shows, as claimed, that we have $17^4 \equiv 1 \pmod{12}$.

Solution 2. The following method involves less multiplication but more thinking. Dangerous curve! ↬ We will make considerable use of this method later in Chapter 18. Dividing 17 by 12 gives a remainder of 5; hence $17 \equiv 5 \pmod{12}$. From equation (7.6) it follows that we have $17^4 \equiv 5^4$. Furthermore, by the Law of Multiplication of Exponents (3.5) we have $5^4 = (5^2)^2$. Since $5^2 = 25$, and since dividing 25 by 12 gives a remainder of 1, we have

$$5^2 \equiv 1 \quad (\text{mod } 12) \tag{7.8}$$

By putting $a = 5^2$ and $b = 1$ in equation (7.7), we obtain

$$(5^2)^2 \equiv 1^2 \quad (\text{mod } 12)$$

Putting the above facts together, we have, as claimed,

$$17^4 \equiv 5^4 \equiv (5^2)^2 \equiv 1 \quad (\text{mod } 12)$$

Example 7.9. Show that $2^{64} \equiv 3 \pmod{13}$.

We will give two solutions to this problem. The first solution uses the method of Solution 1 of Example 7.8, and the second solution uses the method of Solution 2 of Example 7.8.

Solution 1. **The long straightforward solution.** We begin by computing

$$2^{64} = 18446744073709551616$$

After dividing this 20-digit number by 13—we omit the exacting details—the remainder is 3.

Solution 2. **The short clever solution.** This really is the shorter method, although it appears to be longer because we show many more details of this computation than we did for the previous one. This method is like taking apart and putting back together a nested set of Russian babushka dolls.[1] The point of the following transformation of 2^{64} is that the last expression on the right can be evaluated with only 6 multiplications whereas the straightforward method of evaluation requires 63 multiplications.

$$2^{64} = \left(2^{32}\right)^2 = \left(\left(2^{16}\right)^2\right)^2 = \left(\left(\left(2^8\right)^2\right)^2\right)^2$$

$$= \left(\left(\left(\left(2^4\right)^2\right)^2\right)^2\right)^2 = \left(\left(\left(\left(\left(2^2\right)^2\right)^2\right)^2\right)^2\right)^2$$

Using the right side of the preceding equality, we compute the inmost parentheses first and proceed outward. The key is to make multiple use of the Law of Multiplication of Exponents together with equation (7.7). Notice carefully that certain of the following relations are congruences and others are equalities.

$$2^2 = 4$$

$$2^4 = \left(2^2\right)^2 = 4^2 = 16 = 13 + 3 \qquad\qquad \equiv 3 \pmod{13}$$

$$2^8 = \left(2^4\right)^2 \equiv 3^2 \qquad\qquad\qquad\qquad = 9 \pmod{13}$$

$$2^{16} = \left(2^8\right)^2 \equiv 9^2 = 81 = 6 \times 13 + 3 \qquad \equiv 3 \pmod{13}$$

$$2^{32} = \left(2^{16}\right)^2 \equiv 3^2 \qquad\qquad\qquad\qquad = 9 \pmod{13}$$

$$2^{64} = \left(2^{32}\right)^2 \equiv 9^2 = 81 = 6 \times 13 + 3 \qquad \equiv 3 \pmod{13}$$

Why casting out nines works

We are now prepared to see why the curious recipe of casting out nines is a valid method for checking arithmetic.

For the rest of this section, we will use the modulus 9 because—as noted earlier—that modulus relates to casting out nines. Note that every integer is congruent to one of the integers

$$0, 1, 2, 3, 4, 5, 6, 7, 8$$

Casting out nines is a special technique for calculating which one of these nine numbers is congruent to a given arithmetic expression. The first step is the observation that 10 is congruent to 1 modulo 9. From Fact 7.4, it now follows that we have

$$100 \equiv 1 \qquad 1000 \equiv 1 \qquad 10000 \equiv 1 \quad (\text{mod } 9) \qquad\qquad (7.9)$$

and, in fact, for every natural number n,

$$10^n \equiv 1 \quad (\text{mod } 9)$$

This fact together with Fact 7.4 implies the following proposition.

Proposition 7.10. *Let n be a natural number, and let m be the sum of the decimal digits of n. Then n and m have the same remainder under division by 9. In other words, we have*

$$n \equiv m \quad (\text{mod } 9)$$

Instead of giving a formal proof, we verify this proposition in a special case.

Example 7.11. By Fact 7.4 we have

$$
\begin{aligned}
4629 &= 4 \times 10^3 + 6 \times 10^2 + 2 \times 10 + 9 \equiv 4 \times 1^3 + 6 \times 1^2 + 2 \times 1 + 9 \quad (\text{mod } 9) \\
&= 4 + 6 + 2 + 9 = 21
\end{aligned}
$$

In the light of Proposition 7.10, let us reexamine Definition 7.1 for the process of casting out nines. We see (1) that each step of casting out nines from a number n results in a number that is congruent (mod 9) to n, and (2) that the process—illustrated in Example 7.2—terminates in a nonnegative number less than 9 that is congruent (mod 9) to n. Furthermore, we see that the method of casting out nines illustrated in problems (7.1) and (7.2) is validated by repeated applications of Fact 7.4.

The congruence $10 \equiv 1 \pmod 9$ is the only property of the number 9 that is used to prove that the method of casting out nines is valid. Since $10 \equiv 1 \pmod 3$, *casting out threes* is also a valid method. The reason why casting out nines is traditionally the preferred method of checking arithmetic is that the chance of error is smaller. In fact, a random answer to an arithmetic problem has probability $1/9$ of passing the test of casting out nines, whereas the corresponding probability of error is $1/3$ for the method of casting out threes.

Tests for Divisibility

The tests for divisibility of integers can be discussed in the light of modular arithmetic. The following tests for divisibility by 2, 5, 10, and 4 are easy to establish; we will simply state them.

Proposition 7.12. *An integer is divisible by*

- *2 if and only if its last digit is even*

- *5 if and only if its last digit is 5 or 0*

- 10 *if and only if its last digit is* 0

- 4 *if and only if its last two digits represent a number divisible by* 4.

The following is an easy consequence of Proposition 7.10.

Proposition 7.13. *An integer is divisible by* 9 *if and only if the sum of its digits is divisible by* 9.

Divisibility by 3 has a similar test. As in Proposition 7.10, the congruence $10 \equiv 1$ (mod 3) together with Fact 7.4 implies the following.

Fact 7.14. *An integer is divisible by* 3 *if and only if the sum of its digits is divisible by* 3.

As a final example we consider the test for divisibility by 11. This is the only other such test that is useful for hand calculations.

Proposition 7.15. *An integer is divisible by* 11 *if and only if the alternating sum of its decimal digits is divisible by* 11.

Proof. We will show the stronger assertion that an arbitrary integer is congruent modulo 11 to the alternating sum of its decimal digits.

We use the congruence $10 \equiv -1$ (mod 11) together with Fact 7.4. In fact, for an arbitrary integer n, it follows that we have $10^n \equiv (-1)^n$ (mod 11). The right side of this congruence is equal to $+1$ or -1 depending on whether n is even or odd. Suppose an integer has the decimal expansion

$$a_N \cdot 10^N + a_{N-1} \cdot 10^{N-1} + \cdots + a_1 \cdot 10 + a_0$$

Using Fact 7.4 we see that the above expression is congruent modulo 11 to

$$a_N \cdot (-1)^N + a_{N-1} \cdot (-1)^{N-1} + \cdots - a_1 + a_0$$

which is the alternating sum of the decimal digits.   □

> **alternating sum.** The *alternating sum* of a list of numbers
> $$n_1, n_2, n_3, \ldots, n_k$$
> is defined to be
> $$n_1 - n_2 + n_3 - \ldots \pm n_k$$
> The last term gets a plus or minus sign depending on whether there are an odd or an even number of terms.

There are tests for divisibility by 7 or 13, but they are too complicated to be useful pencil-and-paper calculating tools.

We conclude this chapter with an application of congruences that is of great importance in today's computer technology—the generation of random numbers.

Random Numbers

In 1955, a volume was published by the Rand Corporation with the title *One Million Random Digits*.[2] These numbers were obtained from a random physical process, a random frequency pulse source. As noted in the introduction of that work, a book of random numbers does not require careful proofreading to eliminate random errors. Random numbers are useful for computer simulations, also called the Monte

Carlo method. When it is difficult or impossible to calculate theoretically how an airplane wing behaves under all possible conditions, the next best thing is to subject it to Monte Carlo computer testing in which random environmental conditions are represented numerically. In Monte Carlo testing, supercomputers push the limits of computing capability.

Books of random numbers are not published today because it is easy to compute random numbers *with the help of the computer*. In fact, it is so easy to compute random numbers that it would be considered a waste of computer resources to keep a list of them in memory.

Random numbers are not usually generated by a true random process as was done in *One Million Random Digits*. Random numbers are usually computed by a deterministic process, and they are often called pseudo-random numbers. Computer users usually accept the random numbers provided by an operating system, computer language, or program without knowing the details of their construction.

The most widely used random number generators are called *linear congruential generators* (LCG). This method constructs a sequence of random integers $x_0, x_1, x_2 \ldots$ by the following recursive formula:

$$x_{n+1} = ax_n + c \pmod{m} \qquad 0 \le x_n < m \qquad n = 0, 1, 2 \ldots \qquad (7.10)$$

To generate a particular sequence of random numbers, we need to choose integers a, c, and m. In addition, we must choose the first integer of the sequence, x_0, which is called the *seed*. The "random" number sequence generated in this way is actually a deterministic process that is periodic, with a period no greater than m. For the sequence to appear random, the period should be large, and this means that m should be a large number. The choice of the integers a, c, and m is an art. Some choices are much better than others. The appropriate choice depends on the particular use made of the random numbers.[3]

From casting out nines to random number generation, we have seen several applications of congruences. Casting out nines has been used uncritically by thousands who have regarded it as a mysterious dogmatic precept. The insight of this chapter is the realization that this method is a very special application of modular arithmetic. In Chapter 18, we will see the connection between modular arithmetic and public-key cryptography. We have just opened the door to the study of congruences, one of the central topics of the theory of numbers.

We started this chapter by looking at the abacus, an ancient method of computation. We begin the next chapter with another traditional method of computation, Russian peasant multiplication, that leads us to the *binary system*, the arithmetic foundation of the modern digital computer.

Chapter 8

The Power of Two

*And of every living thing of all flesh, two of every sort shalt thou bring into
the ark.*

—GENESIS 6:19

Ten is the base of the decimal system, but any natural number greater than 1 can serve
as a base for representing numbers. The binary system, the base-2 system, is used
in computers. We examine the role of the binary system in a variant multiplication
algorithm, Russian peasant multiplication, and in the analysis of certain games. The
ternary system, the base-3 system, leads us to Cantor's curious middle-thirds set.

The Binary System

Russian peasant multiplication

Russian peasants may not have been mathematicians, but it is said that they used an
interesting alternate method of multiplication.

Example 8.1 (Russian peasant multiplication). Suppose that we want to multiply
57×25. We construct two columns of numbers. The top numbers of the columns are
57 and 25, respectively. In the left column, each number is double the one above it.
In the right column, each number is half the number above it, ignoring any fractional
amounts. The columns end when the number 1 appears in the right column. We
cross out any number in the left column whose right neighbor is even. The product
57×25 is the sum of the numbers in the left column that are not crossed out. This
sum is 1425. By the way, the multiplication $57 \times 25 = 1425$ really is correct!

57	25
~~114~~	12
~~228~~	6
456	3
+ 912	1
1425	

This algorithm for multiplication is based on the base-2 system of representing
numbers, also called the binary system. In fact, we can use any natural number
greater than 1 as a base for a number system. We ordinarily use a base-10 system,
also called the decimal system, for arithmetic. We have just now been writing num-

bers in the 10-system. For example, 1425 is a brief way of writing

$$1 \times 10^3 + 4 \times 10^2 + 2 \times 10 + 5 \qquad (8.1)$$

The method of representing integers in bases other than 10 uses a simple but important property of the integers called the division algorithm. We are not speaking here of the usual arithmetic algorithm of long or short division. The division algorithm states that after dividing a by b one obtains a quotient and a remainder.

> Integers and natural numbers are both whole numbers, but recall that an *integer* can be zero or negative, whereas *natural numbers* must be positive.

Theorem 8.2 (the division algorithm). *Let a and b be natural numbers. Then there exist unique nonnegative integers, the quotient q and the remainder r, such that $a = qb + r$ and $0 \le r < b$.*

For example if $a = 23$ and $b = 7$, then $23 = 3 \cdot 7 + 2$ so that $q = 3$ and $r = 2$. We will not give a proof of this theorem. Our experience with elementary arithmetic makes this theorem plausible.

Note that although this is a theorem concerning division, the arithmetic division symbols ($\div$ or $/$) do not appear. We don't write

$$\frac{a}{b} = q + \frac{r}{b}$$

because this would make it appear that rational numbers are involved. In truth, the division algorithm is concerned solely with the integers.

Before we change decimal numbers into a nondecimal base, we show a simple example using the number 1425 from Example 8.1. It is very easy in the 10-system to divide a number by 10. In this example, we divide 1425 by 10, and then divide the quotient by 10, and so on. At each step we keep track of the remainders of these successive divisions by 10. This arithmetic is trivial, but it is a pattern for more interesting calculations later.

Dividing 1425 by 10 gives a quotient of 142 and a remainder of 5.
Dividing 142 by 10 gives a quotient of 14 and a remainder of 2.
Dividing   14 by 10 gives a quotient of   1 and a remainder of 4.
Dividing   1 by 10 gives a quotient of   0 and a remainder of 1.

Quotients	Remainders
10)1425	*5*
142	*2*
14	*4*
1	*1*
0	

Figure 8.1. Computing base-10 digits.

In the short version of the above calculation, shown in Figure 8.1, we write the successive quotients one under the other and write the remainders in the right column.

Notice that in Figure 8.1 we write 10 to the left of 1425 to indicate that 10 is the divisor for every division in Figure 8.1.

Note that the remainders in the right column of Figure 8.1 are the digits of the number 1425. In fact, we can use the Figure 8.1 to give meaning to a number in the 10-system: 1425 is the number that under successive division, as described above, results in the sequence of remainders shown in Figure 8.1.

A number is uniquely determined by the numbers in the remainders column of Figure 8.1. In the present case, 1425 is the unique number determined by the remainders 1, 4, 2, 5.

The fact that we use powers of 10 is a historical accident. We could equally well use an arithmetic based on powers of 7 or any other natural number greater than 1. Other people at other times have used other bases for numbers. For example, the Babylonians used a base-60 system. We have a remnant of this system in our use of 60 minutes in an hour and 60 seconds in a minute. We use base-10 because from ancient times we have used our 10 fingers for counting. The Mayans used a base-20 system because they lived in a warm climate and had less need to cover their feet for warmth, leaving both fingers and toes available for counting.

The binary system is based on powers of 2 instead of powers of 10. We use a subscript 2 or 10 to specify the base of a number in cases where there might be confusion. In the binary system, numbers are represented using only the digits 0 and 1. To show how Russian peasant multiplication works, we need to transform numbers from our normal number system, the 10-system, to the binary system.

A table like Figure 8.1 can be used to convert numbers from base-10 to some other base. For example, to convert from base-10 to base-2, use 2 instead of 10 as a divisor in a table similar to Figure 8.1. Figure 8.2 converts 25_{10} to base-2. As it was for the 10-system, the digits are in the remainders column with the least significant digit on top and the most significant digit on the bottom.

Quotients	Remainders
2)25	*1*
12	*0*
6	*0*
3	*1*
1	*1*
0	

$$25_{10} = 11001_2$$

Figure 8.2. Computing base-2 digits.

The long version of 11001_2 is

$$25 = 1 \times 2^4 + 1 \times 2^3 + 0 \times 2^2 + 0 \times 2^1 + 1 \times 2^0$$

or, expanding the powers of 2,

$$= 16 + 8 + 1 \tag{8.2}$$

Note that 25 happens to be one of the multipliers in Example 8.1 (Russian peasant multiplication). And the arithmetic that we did just now was the same as part of the calculation in Example 8.1. The rows that were crossed out in Example 8.1 correspond exactly to the digits that are 0's in the present calculation. Thus, we have shown that Russian peasant multiplication computes the binary digits of 25 in Example 8.1. We must still show the function of the successive doubling of the number 57. To show exactly how Russian peasant multiplication works, let us redo Example 8.1 as follows. Note that we are permitted to multiply a number in the decimal system by a number in the binary system. A number is unaffected by our choice to write it in the 10-system, the 2-system, or even in Roman numerals.

Example 8.1 was

$$57 \times 25 = 57_{10} \times 11001_2 \tag{8.3}$$

Using equation (8.2):

$$= 57 \times \left((1 \times 2^4) + (1 \times 2^3) + (0 \times 2^2) + (0 \times 2) + 1 \right)$$

Distributing the multiplication with respect to addition (page 93), and omitting the zero terms:

$$= 57 \times 2^4 + 57 \times 2^3 + 57$$

Expanding the powers of 2:

$$= 57 \times 16 + 57 \times 8 + 57$$

And, finally, evaluating the multiplications:

$$= 912 + 456 + 57 = 1425 \tag{8.4}$$

Russian peasant multiplication does exactly the above operations in a slightly different, more efficient form. The successive divisions in Russian peasant multiplication are used to compute the binary digits of 25. At the same time, the successive doubling of 57 in Russian peasant multiplication is used to compute the multiplication in equation (8.4) of 57 by various powers of 2. The crossed-out terms correspond to the incidence of 0's in 11001_2, the binary representation of 25_{10}. And we see the arithmetic that we did in Example 8.1 is the same as what we just did in equations (8.3) and (8.4).

Russian peasant multiplication is not recommended as a replacement for the usual method of multiplication, but it is interesting to note that calculators and computers use algorithms that are much closer to Russian peasant multiplication than to the paper-and-pencil method that we ordinarily use. Computers convert numbers to the binary system before performing calculations. Computers are intricate systems of switches. These switches have two positions, *on* and *off*, which can be used to represent the binary digits 1 and 0, respectively.

Gottfried Wilhelm von Leibniz (1646–1716), the German mathematician and philosopher, was very impressed that all numbers could be constructed using only 0 and 1. He believed this phenomenon was related to the biblical account of the creation of the universe from the void, and he believed that the universe is constructed of analogues of the number 1 that he called *monads*.

Speculations aside, the binary system appears in many unexpected ways. Readers who have not seen the solution to the following problem will probably find that solution unexpected.

Problem 8.3. Suppose the chef at the student union cafeteria has 10 different spices. In how many different ways can he use these spices to flavor his soup?

People tend to underestimate the number. Actually there are 1024 different combinations of the spices. The binary numbers enter in the following way. We can represent the various spice recipes as 10-digit binary numbers. The first digit tells whether the first spice is used: If the first digit is 1, then the first spice is used in this particular recipe, and if it is 0, then it is not used—and so on for all the nine other digits/spices. For example, we take the number 0110001000 to represent the recipe that uses the second, the third, and the seventh spices and no others. There are 1024 recipes because that is the number of 10-digit binary numbers. (Don't forget to count the number 0000000000—the soup with no spices at all.)

Of course, a similar computation works for any number of spices. Let us state the general result a bit more formally.

Proposition 8.4. *Let S be a nonempty finite set containing n elements where n is any natural number. Then the number of elements in the set of all subsets of S is 2^n.*

The number guessing card trick

The binary system is the basis for a number of games and puzzles. Figure 8.3 is a number guessing card trick. Ask a friend, "Please pick a number between 1 and 31 and don't tell me what it is. Just tell me which cards contain your number." Suppose she responds, "My number is on cards (b) and (d)." Then you respond without hesitation, "Your number is 10." You find the number by adding the numbers in the upper left corners of the selected cards (b) and (d): $8 + 2 = 10$.

It becomes more clear if the numbers on the card are written in the binary as well as the decimal system. Figure 8.5 shows the same cards as Figure 8.3 with the numbers shown in binary as well as decimal notation. In Figure 8.5, we see that the secret of the cards is that card (a) contains all the numbers less than 32 for which the *first* binary digit is a 1, counting from the right. The card (b) has all the numbers less

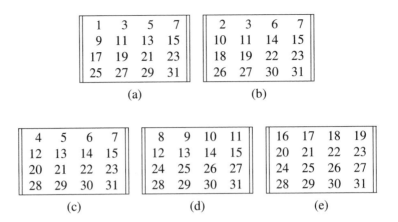

Figure 8.3. Number guessing cards.

than 32 for which the *second* binary digit is a 1, and so on. These digits are shown in italics on the cards. When your friend points to the cards that contain her number, that identifies the binary digits of the chosen number. In the example, cards (b) and (d) were chosen. This reveals that the second and fourth binary digits are 1's and that all the other digits are 0's. In other words, she identified her chosen number as 1010_2, which is the same as 10_{10}.

The game of Nim

The binary system is used in the analysis of the game of Nim[1] that is played as follows. Initially, there are several piles of counters—for example, as shown in Figure 8.4. The two players alternately remove one or more counters. At his turn, a player may remove as many counters as he wishes, but only from one pile. The player who takes the last counter wins. We denote a Nim position by listing the sizes of the piles between curly brackets: $\{6, 7, 13\}$ in Figure 8.4.[2]

Figure 8.4. A Nim position consisting of three piles.

It is unlikely that Nim would be played in a gambling casino—as it was in the film *Last Year at Marienbad* (1961)—because there is a simple key to winning the game. When both players play optimally, the outcome of the game is completely determined by the initial position. In fact, by following the procedure that we will

1 = *1*	3 = 1*1*	5 = 10*1*	7 = 11*1*
9 = 100*1*	11 = 101*1*	13 = 110*1*	15 = 111*1*
17 = 1000*1*	19 = 1001*1*	21 = 1010*1*	23 = 1011*1*
25 = 1100*1*	27 = 1101*1*	29 = 1110*1*	31 = 1111*1*

(a)

2 = *1*0	3 = *1*1	6 = 1*1*0	7 = 1*1*1
10 = 10*1*0	11 = 10*1*1	14 = 11*1*0	15 = 11*1*1
18 = 100*1*0	19 = 100*1*1	22 = 101*1*0	23 = 101*1*1
26 = 110*1*0	27 = 110*1*1	30 = 111*1*0	31 = 111*1*1

(b)

4 = *1*00	5 = *1*01	6 = *1*10	7 = *1*11
12 = 1*1*00	13 = 1*1*01	14 = 1*1*10	15 = 1*1*11
20 = 10*1*00	21 = 10*1*01	22 = 10*1*10	23 = 10*1*11
28 = 11*1*00	29 = 11*1*01	30 = 11*1*10	31 = 11*1*11

(c)

8 = *1*000	9 = *1*001	10 = *1*010	11 = *1*011
12 = *1*100	13 = *1*101	14 = *1*110	15 = *1*111
24 = 1*1*000	25 = 1*1*001	26 = 1*1*010	27 = 1*1*011
28 = 1*1*100	29 = 1*1*101	30 = 1*1*110	31 = 1*1*111

(d)

16 = *1*0000	17 = *1*0001	18 = *1*0010	19 = *1*0011
20 = *1*0100	21 = *1*0101	22 = *1*0110	23 = *1*0111
24 = *1*1000	25 = *1*1001	26 = *1*1010	27 = *1*1011
28 = *1*1100	29 = *1*1101	30 = *1*1110	31 = *1*1111

(e)

Figure 8.5. Number guessing cards with numbers in binary and decimal notation.

describe below, either the first player or the second player can force a win. Which player has the advantage is determined solely by the initial position.

To analyze a Nim position, the first step is to calculate the binary representation of the size of each pile. For the Nim position $\{6, 7, 13\}$ in Figure 8.4, we compute $6 = 110_2$, $7 = 111_2$, and $13 = 1101_2$. Since it is awkward to interrupt the game to find pencil and paper, we will see that this can be done mentally. In each pile we first look for the largest power of 2 that is less than the size of the pile, and we mentally place a marker, an imaginary matchstick, to set apart this many counters. We repeat this process with the remaining counters in each pile—with lower powers of 2, of course—until each pile is dissected into *groups* whose lengths are different powers of 2. For the Nim position $\{6, 7, 13\}$, this grouping is shown in Figure 8.6.

	Powers of 2			
	$2^3 = 8$	$2^2 = 4$	2	1
$6 = 4+2 = 110_2$		▢▢▢▢	▢▢	
$7 = 4+2+1 = 111_2$		▢▢▢▢	▢▢	▢
$13 = 8+4+1 = 1101_2$	▢▢▢▢▢▢▢▢	▢▢▢▢		▢
Number of groups	1	3	2	2
Parity	odd	odd	even	even

Figure 8.6. The position $\{6, 7, 13\}$. Each pile is dissected into groups. This position is unbalanced because some of the parities are odd.

Next, we count the number of groups for each power of 2; we are especially interested in the parity, even or odd, of each of these numbers. For the Nim position $\{6, 7, 13\}$, the parities are shown in Figure 8.6.

Definition 8.5. If a Nim position has only *even* entries in the parity row of Figure 8.6, then we say that the position is *balanced*. Otherwise, we say it is *unbalanced*.

For example, the position $\{6, 7, 13\}$ in Figure 8.4 is unbalanced because in Figure 8.6 there are some odd entries in the parity row.

At his turn, if a player is confronted with an unbalanced position, then he can win by removing counters so that the position becomes balanced. For example, the unbalanced position in Figure 8.4 becomes balanced by replacing the pile of size 13 with a pile consisting of a single counter. This converts the unbalanced position $\{6, 7, 13\}$ into the position $\{6 = 110_2, 7 = 111_2, 1\}$ that, as seen in Figure 8.7, is a balanced position.

We will soon describe the general method for converting an unbalanced position into a balanced position.

On the other hand, at his turn, if a player is confronted with a balanced position, then there is no way for him to force a win.

The important facts about balanced and unbalanced Nim positions are as follows:

	Powers of 2		
	$2^2 = 4$	2	1
$6 = 4 + 2 = 110_2$	☐☐☐☐	☐☐	
$7 = 4 + 2 + 1 = 111_2$	☐☐☐☐	☐☐	☐
1			☐
Number of groups	2	2	2
Parity	even	even	even

Figure 8.7. The position $\{6, 7, 1\}$. This position is balanced because the total number of groups for each power of 2 has even parity.

Proposition 8.6.

1. *A balanced position becomes unbalanced by any play whatever. That is, if a position is balanced, then* every *further single play makes it unbalanced.*

2. *An unbalanced position can always be balanced. That is, if a position is unbalanced, there always exists at least one single play that makes it balanced.*

Before we give the proof, we remark that Proposition 8.6 implies that the best way to play Nim is to observe the following precept.

> **If a position is unbalanced, balance it!**

In fact, suppose that Ada and Ben are playing Nim, and at Ada's turn the position is unbalanced. If Ada follows the precept and balances the position then there are two possibilities: (1) Ada wins immediately because no counters remain or (2) Ada's move causes Ben to be confronted with a balanced position, and Proposition 8.6 asserts that he has no choice but to unbalance it. Proceeding in this way, Ada can be assured that the position is unbalanced (and therefore nonempty) when it is her turn to play. At each play there are fewer counters; therefore, eventually Ada wins by converting an unbalanced position into the balanced position consisting of no counters at all—the empty position. In summary, the unbalanced positions are *winning* positions and the balanced positions are *losing* positions.

Proving 1 of Proposition 8.6 is easy, but to prove 2, it is helpful to introduce some notation and a definition. In Figure 8.8, we amplify Figure 8.6 by using the parities to generate the binary digits of an integer that we call the *parity number* of a position.

Definition 8.7. Let $\mathcal{A}$ be a Nim position. We define the parity number $p(\mathcal{A})$ to be the natural number defined as follows. For each power of 2, the corresponding binary digit of $p(\mathcal{A})$ is 1 or 0 depending on whether the number of the corresponding groups of $\mathcal{A}$ (e.g., as shown in Figure 8.8) has odd or even parity.   ↩ Dangerous curve!

For example, we see that if $\mathcal{A}$ is the position $\{6, 7, 13\}$, then $p(\mathcal{A}) = 1100_2$.

Power of 2	$2^3 = 8$	$2^2 = 4$	$2^1 = 2$	$2^0 = 1$
Number of groups	1	3	2	2
Parity: even or odd	odd	odd	even	even
Binary digit	1	1	0	0
Parity number	$p(\mathcal{A}) = 1100 = 2^3 + 2^2 = 12$			

Figure 8.8. The binary representation of the parity number of the Nim position $\{6, 7, 13\}$ is equal to 1100_2.

Fact 8.8. *A Nim position $\mathcal{A}$ is balanced if and only if the parity number $p(\mathcal{A})$ is equal to 0.*

Proof of Proposition 8.6.

1. Suppose $\mathcal{A}$ is balanced. Then any play alters at least one entry in the parity chart. Since the only possible change is from even to odd, it follows that $\mathcal{A}$ is transformed into an unbalanced position.

Dangerous curve! ↵ **2. On the other hand, suppose that $\mathcal{A}$ is unbalanced.** (For example, in Figure 8.8, we see that the position $\{6, 7, 13\}$ is unbalanced.) Then $p(\mathcal{A})$ is a positive number that we denote n. (For example, in Figure 8.8, $\mathcal{A}$ is the position $\{6, 7, 13\}$, and $p(\mathcal{A}) = 1100_2$.) Put 2^m for the highest power of 2 not exceeding n; in other words, 2^m is the power of 2 corresponding to the leftmost binary digit, the most significant binary digit, of n. (In Figure 8.8, the highest power of 2 not exceeding 1100_2 is $1000_2 = 2^3$; therefore, m is equal to 3.) There must be at least one pile belonging to the position $\mathcal{A}$ that has a 1 for the digit corresponding to 2^m in the binary representation of its size; for the size of this pile we put k. Thus, counting from the right, the $m + 1$st digit of k must be a 1. (In Figure 8.8, the position $\{6, 7, 13\}$ contains a pile of 13 counters and 13 has the binary representation 1101_2; the 1 in the 8's digit, the leftmost digit, agrees with the most significant digit of $n = 1100_2$.)

We will see that the play that transforms $\mathcal{A}$ into a balanced position is one in which counters are removed from this pile of size k. (For the position $\{6, 7, 13\}$ of Figure 8.4, k is equal to 13.) We are permitted to remove any positive number of counters from this pile; in other words, we can replace this pile by any single pile of size j such that $0 \le j < k$. (In Figure 8.8, we can transform the pile of size 13 into a pile of any smaller size, or we can remove the pile of 13 altogether.) The number j is obtained from k by altering some of the binary digits of k as follows.

1. *The $m + 1$st binary digit of k* (counting from the right). This digit, which corresponds to 2^m, is changed from a 1 to a 0.

2. *Binary digits 0 through m of k.* The ith digit ($1 \le i \le m$) is changed (from 0 to 1 or from 1 to 0) if the ith binary digit of the parity number n is a 1.

Notice that 1 and 2 ensure that j is less than k. Furthermore, 1 and 2 guarantee that all the *odd* entries in the parity row (e.g., in Figure 8.8) are changed to *even*. (For the position $\{6, 7, 13\}$ of Figure 8.4, replacing the pile of 13 counters with a pile consisting of 1 counter achieves conditions 1 and 2.)   □

We recapitulate the procedure for winning at the game of Nim. Whenever we make a play in Nim, we seek to leave a balanced position. From Figure 8.6 we see that the position {6, 7, 13} is unbalanced. If it is my move, I can win by making it balanced by replacing the pile of 13 with a pile consisting of 1 counter. In other words, the winning move converts the unbalanced position {6, 7, 13} into the balanced position {6, 7, 1}.

Here is the step-by-step procedure.

1. Make a table like Figure 8.6 showing the parities of the various groups of powers of 2.

2. Identify the largest power of 2 that has odd parity. In Figure 8.4, this power of 2 is $8 = 2^3$. Here this happens to also be the largest power of 2 in any group, but in general this might not be the case.

3. Find a pile of size not less than the power of 2 ($8 = 2^3$) identified above in 2. In Figure 8.4, there is only one pile with 8 or more counters, the pile of 13 counters.

4. Remove counters from the pile determined in 3 in such a way that even parity is restored to every power of 2. In Figure 8.4, the winning play is to remove all but one of the counters from the pile of 13 counters. It is easy to check that all the parities then become even.

The game of Kayles

Kayles[3] is similar to Nim, except that the piles of counters must be arranged in rows, and instead of taking any number from any single row, the rule in Kayles is that one must take one or two adjacent counters from a row. Removing counters from the middle of a row changes the original row into two smaller rows. Just as in Nim, the person who takes the last counter wins.

Here is an example of a game played by our imaginary players, Ada and Ben. The initial configuration is arbitrary; for this example suppose that it is the following:

□□□□□□□   □□□□□□

Ada starts by taking two counters.

□□□□□□□   □   □□□

Ben responds by taking one counter.

□□□□□□   □   □□□

Ada takes one.

□□□□□   □□□

Ben takes two.

⬜⬜   ⬜⬜   ⬜⬜⬜

Ada takes one.

⬜⬜   ⬜⬜   ⬜   ⬜

Now Ada has won because there are two 2's and two 1's. Ada can simply mimic any move that Ben makes until all the counters are gone.

Perhaps Ada knows the secret of the game of Kayles. If so, she is familiar with Figure 8.9—the Rosetta stone[4] that translates Kayles rows into Nim piles.

We obtain a Nim position equivalent to a given Kayles position by replacing each Kayles row with the Nim pile specified in Figure 8.9. A Kayles position is a winning or losing position depending on whether the equivalent Nim position is unbalanced or balanced. Since there is an arithmetic procedure for determining whether a Nim position is unbalanced or balanced, this equivalence enables us to determine whether a Kayles position is winning or losing.

Notice that the last two N-columns of Figure 8.9 are identical. In fact, all the N-columns are identical apart from the 13 exceptional values that are shown in underlined italics. The table stops at $n = 96$, but the further values of N repeat exactly with the period 12—there are no further exceptions.

The balanced/unbalanced concepts of Nim are used in Kayles, but not in the same straightforward way because one must memorize Figure 8.9. This table gives N as a function of K. (For a discussion of functions, see page 46.)

K	N	K	N	K	N	K	N	K	N	K	N	K	N	K	N
1	1	13	1	25	1	37	1	49	1	61	1	73	1	85	1
2	2	14	2	26	2	38	2	50	2	62	2	74	2	86	2
3	*3*	15	*7*	27	8	39	*3*	51	8	63	8	75	8	87	8
4	1	16	1	28	*5*	40	1	52	1	64	1	76	1	88	1
5	4	17	4	29	4	41	4	53	4	65	4	77	4	89	4
6	*3*	18	*3*	30	7	42	7	54	7	66	7	78	7	90	7
7	2	19	2	31	2	43	2	55	2	67	2	79	2	91	2
8	1	20	1	32	1	44	1	56	1	68	1	80	1	92	1
9	*4*	21	*4*	33	8	45	8	57	*4*	69	8	81	8	93	8
10	2	22	*6*	34	*6*	46	2	58	2	70	*6*	82	2	94	2
11	*6*	23	7	35	7	47	7	59	7	71	7	83	7	95	7
12	4	24	4	36	4	48	4	60	4	72	4	84	4	96	4

Figure 8.9. Translation from Kayles to Nim. The K-columns are the natural numbers in their normal order; we interpret K as the length of a single row of counters in Kayles. In the table, K is the length of a Kayles row; the corresponding number N is the size of the equivalent Nim pile. The N-value corresponding to a natural number K is denoted $N(K)$ and is called the *Nim equivalent* of K.

Figure 8.9, which enables one to translate a Kayles position into an equivalent Nim position, appears to be too much to memorize, but actually it isn't because the table is very repetitive.

As in Nim, one wins in Kayles by playing so as to achieve balanced positions. A balanced Kayles position is one whose equivalent Nim position is balanced. We determine an equivalent Nim position by using Figure 8.9 as follows.

Let us differentiate between Nim positions and Kayles positions by using curly brackets for Nim positions and square brackets for Kayles positions. For example, in the game of Kayles that we just played, the initial position was $[7,6]$. To obtain the equivalent Nim position, replace each Kayles row by a Nim pile whose size is equal to the N-value of the Kayles row. In particular, the Kayles position $[7,6]$ is equivalent to the Nim position $\{N(7), N(6)\}$, and by looking in the table we see that this is $\{2,3\}$, an unbalanced Nim position. We see that it is unbalanced by expressing each of these numbers as a sum of powers of 2:

$$\{2 = 2^1, 3 = 2^1 + 2^0\}$$

There are two 2^1's, an even parity, but there is one 2^0, an odd parity. Since not all the parities are even, the Nim position $\{2,3\}$ is unbalanced, and therefore also the Kayles position $[7,3]$ is unbalanced. The play that balances the Nim position consists in taking 1 away from the pile of 3, resulting in the balanced position $\{2,2\}$. This means that it is also possible to balance the Kayles position by altering the second row, the row of 6 counters. In fact, by removing 2 counters from the row of 6 we can create a row of 1 and a row of 3 that results in the Kayles position $[1,3,7]$. Using Figure 8.9 we see that $[1,3,7]$ is equivalent to

$$\{N(1), N(3), N(7)\} = \{1,3,2\} = \{1, 1+2, 2\}$$

which is balanced. Therefore, $\{1,3,7\}$ is also balanced, and this shows that we have found the right play.

> Q. I see that this play is the correct one, the one that balances the Kayles position. I see why you made your play on that particular row, but how did you find that play from among all the other plays involving that row?
>
> A. That part was trial and error. There are not many plays involving that particular row, but I know that one of them creates balance. I just have to analyze the possibilities using Figure 8.9 in the manner that I have just shown. Sooner or later I find the correct play in this way.

Figure 8.9 defines the equivalent Nim positions for Kayles positions consisting of single rows. Figure 8.9 was constructed so that the function $N(K)$, which defines the equivalent Nim positions, satisfies the following two properties. Let K be a natural number.

1. If m is any one of the numbers $0, 1, \cdots, N(K) - 1$, there is a legal Kayles play that transforms the one-row Kayles position $[n]$ into a Kayles position that is equivalent to the one-pile Nim position $\{m\}$.

2. There is no legal Kayles play that transforms the Kayles position $[n]$ into a Kayles position that is equivalent to the one-pile Nim position $\{N(K)\}$.

> Q. Why is the table periodic apart from a few exceptions?
>
> A. No one knows. However, one can prove that Figure 8.9 is eventually periodic. Other similar games have quite different periodicities, and some don't appear to have any periodicity. This question has been studied, but no one has a good answer to this question.

You see that with Figure 8.9 it is almost as easy to win at Kayles as at Nim. It is fairly easy to find the correct play, that is, the play that converts an unbalanced position into a balanced position.

The Ternary System

> One for my Master, one for my Dame,
> And one for the Little Boy that lives in the lane.
>
> —MOTHER GOOSE

The base-3 system of representing numbers is called the *ternary system*. We will see a card trick based on the ternary system, but first let's recall what it means to represent numbers in the ternary system. The ternary system represents numbers using just three digits: 0, 1, and 2. For example, the ternary representation 10212_3 represents the number

$$1 \times 3^4 + 0 \times 3^3 + 2 \times 3^2 + 1 \times 3^1 + 2 \times 3^0$$

which is equal to

$$81 + 0 + 18 + 3 + 2 = 104$$

The middle-thirds card trick

Here is a trick using 27 cards from an ordinary deck of playing cards.

1. Ask a friend to think of one of the 27 cards without telling you which one.

2. Shuffle the 27 cards.

3. Deal the cards face up into three rows of nine cards each as in Figure 8.10.

   We number the cards starting with 0 instead of 1 for reasons that will become clear later. Of course, these numbers don't actually appear on the playing cards. They are mental constructions that help us understand the trick.

4. Ask your friend which of the three rows contains her chosen card.

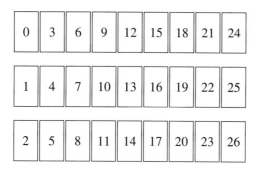

Figure 8.10. The numbers of the cards indicate the order in which the cards are dealt beginning with the top of the deck.

5. Form three piles from the three rows of cards, maintaining the order of the cards within each row.

6. Form a deck by placing the three piles together, making sure that the pile containing the chosen card is the middle pile.

7. Repeat steps 3 to 6 two more times.

8. The chosen card is now exactly in the middle of the deck. Show your friend her chosen card by counting to the fourteenth card in the deck, the card that bears the number 13 above at step 3.

Why does the card trick work?

Notice that the chosen card must belong to the middle third of the deck at the first iteration of step 6. But what if the chosen card is *initially* in the middle third of the deck? In that case, at step 3 the chosen card belongs to one of the three middle *columns* of Figure 8.10. At step 6, the chosen card is in the middle third of its pile and this pile is placed between the two other piles. At this point, the chosen card is in the middle third of the middle third, that is, the middle ninth, of the entire deck. In other words, at the next iteration of step 3 the chosen card has a number in Figure 8.10 between 12 and 14, inclusive. A similar argument shows that if the chosen card is *initially* in the middle ninth of the deck, then at the next iteration of step 6, the chosen card is in the middle third of the middle ninth, that is the middle twenty-seventh of the deck. But the middle twenty-seventh of the deck consists of *exactly one card*, the card bearing the number 13 in Figure 8.10, at the next iteration of step 3.

The foregoing argument shows that at the first iteration of step 6, the chosen card belongs to the middle third of the deck; at the second iteration, it belongs to the middle ninth; and at the third iteration of step 6, it belongs to the middle twenty-seventh of the deck. But the middle twenty-seventh consists of just one card, the fourteenth card, the card bearing the number 13 in Figure 8.10.

The middle-thirds card trick utilizes the ternary system. At the first iteration of step 3, we do not know the number of the chosen card. This number has a certain ternary representation: XXX_3. (With this notation, we do not mean to say that the three digits are the same, merely that the digits are unknown. Of course, each digit is equal to 0, 1, or 2.)

At the second iteration of step 3, we know that the chosen card bears a number of the form $1XX_3$ because at the previous iteration of step 6 we placed the chosen card in the middle third of the deck. In other words, the chosen card now bears a number between 100_3 and 122_3, inclusive, that is between 9 and 17.

At the third and final iteration of step 3, the chosen card bears a number of the form $11X_3$, a number between 110_3 and 112_3, that is between 12 and 14. At the final iteration of step 6, the chosen card has the number 111_3; in other words, the number 13, which makes it the fourteenth card in the deck.

> Q. The ternary system seems to be much less important than the binary system. You have said that the entire technology of computers is based on the binary system, yet for the ternary system all you offer is a card trick. Are there any serious applications for the ternary system?
>
> A. From a certain point of view, the ternary system is just as important as the 10-system because the 10-system represents an arbitrary choice. If we were three-toed sloths instead of humans with 10 fingers we might use the ternary system instead of the 10-system. However, in pure mathematics there is an interesting and important application of the ternary system, Cantor's middle-thirds set.

Cantor's middle-thirds set

German mathematician Georg Cantor (1845–1918) devised a remarkable set of real numbers. The shortest way to define this set is by using the ternary representation of numbers. Every real number can be represented as an infinite ternary expansion. For example, the ternary expansion

$$(.0101010101\ldots)_3$$

represents the number

$$\frac{0}{3} + \frac{1}{3^2} + \frac{0}{3^3} + \frac{1}{3^4} + \frac{0}{3^5} + \frac{1}{3^6} + \cdots$$

Using the formula for the sum of an infinite geometric progression,[5] we can see that this number is actually equal to $1/8$.

We now state the following definition.

Definition 8.9. The *middle-thirds set* of Cantor consists of all real numbers between 0 and 1 (inclusive) that have *at least one* ternary expansion that does not contain the digit 1.

Q. Why do you say, "at least one"? Doesn't every real number have exactly one *ternary expansion?*

A. No, on the contrary, every real number with a terminating *ternary expansion has a second ternary expansion. For example, the number $1/3$ has the terminating ternary expansion $.1_3$ and also a second ternary expansion*

$$(.02222\ldots)_3$$

This means that $1/3$ belongs to Cantor's middle-thirds set. On other hand, $4/9$ does not belong because its terminating expansion $.11_3$ and its alternate ternary expansion

$$(.102222\ldots)_3$$

both contain the digit 1.

In an earlier chapter, for certain purposes we disallowed decimal expansions ending in an infinite repetition of the digit 9. However, it suits our present purposes to allow a second ternary expansion.

There is a way of constructing Cantor's set that makes it easier to visualize. Let's start with the entire line segment from 0 to 1 which we will represent as follows:

0   1

Now divide this interval into thirds. Remove the middle third of the interval $[0, 1]$. The result is the following two subintervals:

0   1

Then remove the middle third of these two subintervals which leaves the following four subintervals:

0   1

Repeat the process:

0   1

And again:

0   1

Again:

0   1

And so on.

Of course, we don't have time actually to carry out the process *infinitely* many times. A point on the line segment $[0, 1]$ belongs to the set unless it is removed by *finitely* many applications of this method.

The following argument shows that this set is very sparse. Suppose that we keep track of the total length of all the subintervals that remain after carrying out this process a number of times. In the first step we removed one-third of the points, and this means that the total length of the remaining intervals is two-thirds. At the second step, the total length of the remaining points is two-thirds of two-thirds, $(2/3)^2$, which is equal to $4/9$. At the nth step the total length of the remaining intervals is $(2/3)^n$. But $(2/3)^n$ becomes as small as we wish if we take n large enough.

> *Q. After so much is removed, does anything remain?*
>
> *A. Clearly, the number 0 is never removed. In fact, the numbers that are never removed are precisely those described in Definition 8.9—a fact that we will not prove. Cantor's set doesn't contain any subintervals, and therefore it doesn't make sense to talk about the total length in the usual sense. Nevertheless, it is possible to generalize the idea of length to something called* measure, *and it makes sense to talk about the measure of Cantor's middle-thirds set. It would take us too far afield to define measure. Suffice it to say that the fact that Cantor's set is contained in a system of intervals of arbitrarily small total length shows that Cantor's set has measure 0.*

Cantor's set has many strange properties. We have shown that Cantor's set is in a certain sense very small. However, in another sense, which we will discuss later, Cantor's set is very large. This paradoxical property is one of the reasons that Cantor's set is interesting to mathematicians.

> *Q. Is Cantor's set important in technology?*
>
> *A. Cantor's set has been suggested as a mathematical model for the occurrence of errors in electronic transmission lines.[6] Cantor's set is important in mathematics because it shows strange, unexpected implications of straightforward plausible definitions.*

There is one more sense in which Cantor's set is very small. It is an example of a *nowhere dense* set, which means that any subinterval of $[0, 1]$, however small, contains a subinterval that contains no points of Cantor's set.

The Sierpinski triangle, a fractal

The name *fractal* was coined in 1975 by the mathematician Benoit Mandelbrot (1924 –) to describe geometric shapes that have a similar appearance at any degree of magnification. Cantor's set is an example of a *fractal* because a triple magnification of the portion of Cantor's set between the numbers 0 and 1/3 is identical to the entire Cantor's set.

The word *fractal* is not defined with the same precision as *circle* or *ellipse*. Fractal is a more loosely defined word like *rough* or *smooth*. There are objects in nature that are considered fractals. For example, a map of the coastline of Britain has a similar jagged appearance at any magnification. A glacial ice fall consists of a jumble of ice blocks from very large to very small, but all of similar shape. An ice fall is an example of a fractal landscape because it tends to look the same under a wide range of magnifications. A picture of an ice fall gives us no clue of the scale of the scene until we see a climber standing beside an ice block the size of an apartment house. Some ferns have a branching structure that gives them a fractal appearance because, under magnification, the smallest sub-sub-branches appear similar to the large main branches.

The Sierpinski triangle (Figure 8.11) was introduced in 1916 by the Polish mathematician Waclaw Sierpinski (1882–1969). This figure can be considered a two-dimensional analogue of Cantor's set. In Figure 8.11, the initial figure is an *equilateral* triangle, but the following construction applies equally well to any triangle.

1. Divide the triangle into four congruent subtriangles by drawing line segments connecting the midpoints of the three sides.

2. Remove the middle triangle.

3. Repeat 1 and 2 and 3 for each of the three remaining triangles.

Note that this process is recursive and never ending because step 3 requires three more repetitions of step 3, and so on. The residual set, after infinitely many smaller and smaller triangles are removed, is the Sierpinski triangle. Like the construction of Cantor's set, this is an infinite process.

Suppose that the initial large triangle has area 1. Let us determine how much is removed at each iteration .

1. A triangle of area $1/4$ is removed.

2. Three triangles each of area $1/4^2$ are removed, a total of $3/4^2$.

Figure 8.11. The Sierpinski triangle.

3. Nine (3^2) triangles each of area $1/4^3$ are removed, a total of $3^2/4^3$.

4. Twenty-seven (3^3) triangles each of area $1/4^4$ are removed, a total of $3^3/4^4$.

The total area removed is

$$\frac{1}{4} + \frac{3}{4^2} + \frac{3^2}{4^3} + \frac{3^3}{4^4} \cdots$$

Using the formula for the sum of n terms of a geometric progression, we find that the total area removed in n steps is

$$\frac{1}{4} \cdot \frac{1 - \frac{3^n}{4^n}}{1 - \frac{3}{4}} = 1 - \left(\frac{3}{4}\right)^n$$

Since the right side of this equation tends to 1, the total of the areas of the removed triangles is 1. As with Cantor's set, we express this fact by saying that the Sierpinski triangle has measure 0.

The Sierpinski triangle is a fractal because the entire triangle is identical to a two-fold magnification of each of three subtriangles.

The Chaos Game

The Chaos Game[7] involves the Sierpinski triangle. The play proceeds as follows as shown in Figure 8.12.

1. Paint the three vertexes of a triangle red, blue, and green, respectively.

2. Paint two sides of a die red, two blue, and two green.

3. Designate an arbitrary point in the triangle as the initial dot ("start" in Figure 8.12).

4. Throw the die.

5. Place a dot on the straight line from the initial point to the vertex with the color shown by the die at a distance halfway to the vertex.

6. After repeating this process about six times, mark the successive locations of the dots.

7. After a large number of repetitions, the marked dots will outline the Sierpinski triangle.[8]

The reason why the marked dots outline the Sierpinski triangle is as follows. Let $\mathcal{T}_1$ be the first triangle removed in the construction of the Sierpinski triangle; let $\mathcal{T}_2$ be the set of three triangles removed in the second step; let $\mathcal{T}_3$ be the set of nine triangles removed in the third step; and so on. If the point "start" belongs to a triangle in the set $\mathcal{T}_n$, then moving the point halfway to a red, blue, or green vertex puts the point in a triangle of the set $\mathcal{T}_{n+1}$. Note that the triangles in $\mathcal{T}_{n+1}$ are half

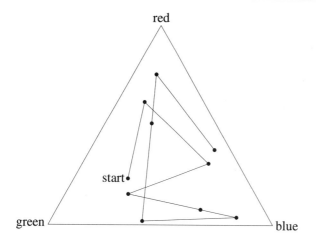

Figure 8.12. The Chaos Game. Designate an arbitrary point in the triangle as "start." Throw a die with two sides painted red, two painted blue, and two painted green. Advance the point half of the distance toward the vertex of the color shown by the die. After this has been repeated about six times, mark each further location of the point. After a large number of throws, the marked points will outline the Sierpinski triangle. The path shown is the result of throwing the sequence: red, blue, green, blue, blue, green, red, red.

the size of the triangles in $\mathcal{T}_n$. The point moves into smaller and smaller triangles. As we see in Figure 8.11, the points in these small triangles are eventually very close—indistinguishably close—to the residual points, the points that belong to the Sierpinski triangle.

In this chapter, we have seen applications of the binary and ternary systems.[9] In the next chapter, we will discuss an application of the binary system that changed our world, the digital computer.

Chapter 9

Divide and Conquer

United we stand, divided we fall.
　　　　　　　　　—ÆSOP (c. 620 – c. 560 B.C.), *The Four Oxen and the Lion*

We will consider a few aspects of the computer, but we do not intend to give a complete introduction to computers. Our goal is the discussion of algorithms that are used on the computer for sorting a list of items alphabetically or numerically. We will look at these algorithms as they might be applied to sort playing cards by hand. The most elegant and powerful of the sorting algorithms employ a technique known as *divide and conquer*.

Computers

Digital computers are huge assemblages of interdependent on-off switches. Today's computers are electronic, but more than 100 years ago, Charles Babbage (1792–1871) designed a mechanical digital computer; the future may bring still another technology. The state (on-off) of each switch is determined by the states of other switches in the computer or by external input devices. The switch settings are used to represent numbers as sequences of binary digits, and arithmetic operations can be carried out by arranging suitable connections and dependencies among the switches. Computers carry out an extensive computation by following a sequence of coded instructions. Prior to 1945, computers received instructions from an external source such as paper tape or punched cards, but a crucial advance occurred when mathematician John von Neumann (1903–57) conceived of a digital computer with a flexible internally stored program of instructions. The internal program is flexible because any desired sequence of instructions can be entered into the computer. The computer then interprets a particular configuration of switches, as needed, either as a number or as an instruction.

　　If a computer merely read and executed a list of arithmetic instructions, it would be useful, but it would not be the powerful device that we know today. The computer derives its great power from *branching* instructions. For example, there is an instruction that tells the computer to find its next instruction at one of two different locations depending on whether a number at a particular location is zero. This enables the computer to repeat a certain subsequence of operations a specified number of times or until a certain condition is met; such a repetitive process is called a *loop*— a *for* loop (*for x equal* 1 *to* 10, *do something*) or a *while* loop (*while x is greater than zero, do something*).

Computer Languages

Numerically coded computer instructions are called *machine language*. Each type of computer has its machine language, and each machine language has its *assembly language* that replaces the numerical codes of machine language with mnemonic translations. For example, in some assembly languages, the code word *mov* stands for a numerical instruction that moves a number from one location to another.

During 1954–58, a computer language called FORTRAN (FORmula TRANslation) was developed which is more than just a word-for-word translation of machine language. FORTRAN contains constructs involving many machine instructions and is designed especially for the needs of scientific and engineering computations. FORTRAN has gone through many refinements and remains the dominant computer language for science and engineering despite certain shortcomings.

Assembly language is a *low-level* language whereas FORTRAN is a higher level language that provides a buffer between the programmer and the more arcane machine and assembly languages.

Higher level computer languages supply abstractions that make computer programming more understandable and error free. In particular, they permit the programmer to give convenient mnemonic names, called variables, to program elements. Moreover, they attend invisibly to many routine "housekeeping" tasks. However, in applications for devices with very limited computing resources (e.g., watches and automobile fuel injection systems), it is necessary to use assembly language to program "on the bare metal" to achieve the maximum speed and greatest control in minimum space.

In FORTRAN, the instruction GOTO—the instruction that passes control to another instruction at a specified location—is a source of difficulty because injudicious use of this instruction can lead to a program that is extremely difficult to understand. GOTO statements can make the flow of control so complex that the result is a tangled mess called a "spaghetti program." This is a serious problem in writing and maintaining large programs—especially when one must understand code created by another programmer who might not be currently involved in the project.

Soon after the introduction of FORTRAN, an effort was under way to produce the language ALGOL (ALGOrithmic Language) the first *structured* computer language. Structured languages are characterized by the following features.

- Structured languages facilitate "top-down" programming. This means that the program is organized in the form of an outline with blocks of instructions located under headings and subheadings. The outline structure is shown by means of indentation. Stepwise refinement divides the main problem into successively smaller problems. (We will soon see a more specific meaning of "divide and conquer.")

- Structured languages simplify the flow of control. The infamous GOTO is available, but is seldom needed because subtasks are performed by blocks of code called *procedures* that are invoked by name rather than by location.

- Structured languages limit the *scope* of program elements. For example, if the programmer defines a variable or procedure, she can restrict the parts of the program in which these structures are meaningful. This means that if the programmer or a co-worker either inadvertently or intentionally uses the same name elsewhere in the program, no clash need occur.

Today ALGOL is supplanted by other structured languages—for example, Pascal, which was designed 1967–71 by Niklaus Wirth.

Pseudocode

Before coding a complicated algorithm in a standard computer language, programmers frequently write out the process in an informal language called *pseudocode* that is part ordinary English and part the Pascal computer language—we might call it pidgin Pascal. Fortunately, anyone who understands English can understand pseudocode. Ordinarily, pseudocode is used for organizing a computer program, but it could also describe culinary tasks.

```
procedure make jelly sandwiches
    begin
        if bread not sliced
            then slice bread
        open jelly jar
        open butter dish
        for each sandwich needed
            butter top slice
            jelly bottom slice
            place top slice on bottom slice
        close jelly jar
        close butter dish
    end
```

Sorting

There are two aspects of computer science, the hardware—the physical devices—and the software—the programs. The analysis of algorithms is an important software concern; we will consider the particular class of algorithms that sort items alphabetically or numerically. The computer demands careful study of sorting algorithms

The reader who uses a computer spreadsheet can see a remarkable demonstration of almost instantaneous sorting. The details of computation are invisible to the user; however, most computer spreadsheets use a divide-and-conquer algorithm.

because (1) the computer, with its great speed, is called on to perform very large-scale sorting for which algorithmic efficiency is crucial, and (2) the computer has abilities that are, not just quantitatively, but also qualitatively different from human beings. We begin with a familiar example of sorting by hand on a small scale.

Sorting playing cards

The methods by which we sort playing cards exemplify the methods that can also be used by computers. We use the sorting of playing cards as a case study to illustrate the analysis of algorithms. When we sort playing cards, we can use more than one method. The most commonly used methods are *insertion sort* and *selection sort*. We say that one card is *higher* than another if the one comes before the other in the sorting order, and we always sort a hand of cards so that the highest card is on the left. Although there are many other sorting methods, we discuss the following three methods and consider their suitability for sorting playing cards by hand.

- **Insertion sort.** We examine each card as it is dealt, and, one by one, we place each card in its proper place.

- **Selection sort.** We look at an entire hand of cards in the random order in which they are dealt. We take the leftmost card as the provisional highest card. Then we examine the next card to the right. If it is larger, then we note that it becomes the provisional highest. We continue to look card by card to the right. The card that is the provisional highest at the end of this process is the true highest, and we move this card to the leftmost position. Now we repeat the process as follows. The leftmost unsorted card is now the provisional second-highest card. Continuing card by card to the right, if we encounter a card higher than the provisional second-highest card, then that card becomes the provisional second highest. On finding the true second highest card, we move it to the immediate right of the highest card. Similarly, we find and move the third-highest card, and so on.

There are many other sorting methods. *Bubble sort* is an algorithm that is sometimes implemented for computer sorting (although there are faster methods), but we never see it used at the card table. Here is how it would work.

- **Bubble sort.** We examine the entire hand of cards in the random order in which it is dealt. If the leftmost card is lower than the card next to it, then we swap these two cards so that the higher of the two is leftmost. Now examine the second card. (It might be the card that we just moved.) If it is lower than the third card, then swap those two cards. Continue through the entire hand of cards in this way. Repeat the process until the cards are finally in the correct order.

  The name bubble sort comes from the fact that the high-ranking items rise like bubbles in champagne.

Why does bubble sort seem inappropriate for sorting cards by hand? It is because, for small-scale hand sorting, bubble sort takes longer than insertion sort or selection sort. Let us examine why bubble sort could be appropriate for computer sorting, but not for sorting playing cards by hand.

Running time

By *running time* we mean the time necessary to carry out an algorithm. The term suggests that a machine of some kind must be involved, but running time analysis is equally meaningful for the methods for sorting playing cards by hand because the discussion of running time is concerned with the number and kind of operations necessary to perform the algorithm rather than the speed of the device that carries out the computation. For arithmetic algorithms, we are interested in the number of multiplications or other arithmetic operations that are required. We count first the operations that consume the most time. For example, since addition is much faster than multiplication, we give more importance to counting the number of multiplications.

In sorting algorithms there are no arithmetic operations. The two operations that are needed for sorting are *data comparison* and *data movement*. All sorting algorithms use multiple instances of these operations.

- **Data comparison.** For the computer, a comparison consists in noting which of two numbers is larger. In card sorting, a comparison consists of determining which of two cards is higher. In sorting cards by hand, this operation is performed by glancing at the cards and mentally deciding which card is higher. Since no physical movement is involved, it is much faster to compare two cards than it is to move a card to a different location. In insertion sort, when we glance at our hand of cards to decide where to put a certain card, we make a data comparison between this card and every card in our hand starting at the left and continuing to the right until we reach the insertion point. Since, roughly, one can make a comparison of ten cards in about 0.1 second, one comparison might take about 0.01 second.

- **Data movement.** The computer moves data by placing it in a different location in its memory. In sorting cards by hand, we move a card physically from one place to another. Very roughly, moving one card takes about 0.5 second.

For sorting cards by hand, it appears that comparison is about 50 times as fast as movement of a card. On the other hand, for computers, data comparison and data movement are roughly equally time consuming. For today's desktop computers, a single data comparison or a single data move takes less than a millionth of a second.

However, in general, we should not ignore the running time of an operation even if it is fast. There may be so many iterations required of the faster operation that it dominates the running time.

Counting data moves

Let us count the number of data moves for these three methods of sorting cards.[1] Let us assume that there are n cards to be sorted.

1. For insertion sort, each card is moved exactly once; hence, whatever the initial order, there are always n moves.

2. For selection sort, the worst case is that the cards are initially dealt ordered from lowest to highest instead of highest to lowest; in this case there are $n-1$ data moves because every card, except one, must be moved. (Note that after $n-1$ moves the last card is in the right place without any need to move it.) In the best case, the cards are initially in the correct order and no moves are needed.

3. In bubble sort, the worst case, as in selection sort, is that the cards are in exactly the wrong order from lowest to highest. On the first pass, the lowest card is moved $n-1$ times from the leftmost to the rightmost position. On the second pass, the second lowest card is moved $n-2$ times down to its proper place; and so on. The total number of moves is the sum of the first $n-1$ natural numbers. In Chapter 5, we called this sum T_{n-1}, and we found that it is equal to $n(n-1)/2$.

We see that, of these three methods, bubble sort requires the most data moves for large n. Since data moves are slow for hand sorting, this explains why bubble sort is not a good method for sorting playing cards.

Order of magnitude

In the analysis of running time, rough estimates are sufficient because we are mainly interested in whether a computation is feasible. Generally, there is a number, in our case n, the number of items, that describes the size of the problem. In computer science, we are interested in the running time of algorithms mainly when n is large because we want to determine how large a problem the computer can handle and how long it will take. We want to know how running time grows as n becomes large. For this purpose, *order-of-magnitude* estimates are sufficient. The following definition gives a technical meaning to order of magnitude.

For the formula

$$f(n) = O\{g(n)\}$$

we say, "f of n has *order g of n*." to emphasize the fact that we are talking about an upper bound, we could say, "f of n has order g of n *at most*."

Definition 9.1. Let the nonnegative functions $f(n)$ and $g(n)$ be defined for all natural numbers n. If there exists a constant c such that $f(n) < cg(n)$ for all sufficiently large natural numbers n, then we say that, as n tends to infinity, the function $f(n)$ has order of magnitude $g(n)$, and we write

$$f(n) = O\{g(n)\}$$

Note that if $f(n) = O\{g(n)\}$, then $f(n) = O\{h(n)\}$ for any function $h(n)$ such that $h(n) \geq g(n)$ for all sufficiently large n.

Q. Why is there a "<" in this definition? What good is it to know that the running time is less than something?

A. We are interested in an upper bound, preferably small, for the running time because then we can confidently assert that the algorithm can be performed on a particular computer within a certain time. Our first concern is the feasibility of the computation.

Q. But what good is it to know that the running time is less than something times a constant c, when we don't know whether the constant is equal to 1 or 1,000,000?

A. We do not need to know the constant c to determine the percent growth of running time as n increases. For example, suppose that running time is $O\{n^2\}$; in this case, if we double n, then running time—more accurately, the upper bound on running time—quadruples. For certain tasks, more complex than sorting, the best available algorithms have order-of-magnitude estimates that are exponential in n, for example, $O\{2^n\}$. This is considered very unsatisfactory, because simply adding 1 to n tends to double the running time. An algorithm with order of magnitude $O\{n^{100}\}$ is faster for large values of n than one with an exponential order of magnitude. (See the last row of Figure 9.1.)

Q. If we are told that running time is $O\{n^2\}$, how can we be sure that running time might not have a much smaller order of magnitude— for example, $O\{n\}$?

A. That is logically possible. However, the accepted practice is to state the smallest known order of magnitude. Generally, one would not state the order of magnitude $O\{n^2\}$ if $O\{n\}$ were known.

Notation. The formula $\log n$ means "logarithm to the base 10 of n." By definition, $\log n$ is the exponent m such that $10^m = n$, but in order-of-magnitude statements such as $O\{\log n\}$ or $O\{n\log n\}$, it is close enough to approximate $\log n$ as the number of decimal digits in the number n. If convenient, we can count the binary digits instead of the decimal digits; this latitude is possible because the value of the constant c in Definition 9.1 can be adjusted suitably. (The number of binary digits is approximately equal to 3.3 times the number of decimal digits.)

In sorting algorithms, $g(n)$ is usually one of the three functions: n, $n\log n$, or n^2. Sorting algorithms are all rather fast; there are other computational list-processing tasks that are so difficult that the fastest known algorithms have a running time that is exponential, for example $O\{2^n\}$, where n is the number of items. An exponential running time is worse than $O\{n^{100}\}$ because for n sufficiently large, $2^n > n^{100}$. In fact, $O\{2^n\}$ is worse than $O\{n^k\}$ for any k, however large. Figure 9.1 exhibits the growth of a few functions that could be used in order-of-magnitude statements.

Example 9.2. The number of data moves for insertion sort is $O\{n\}$ where n is the number of items to be sorted.

Proof. Definition 9.1 requires us to show that for some constant c, the number of moves is less than cn for all sufficiently large n. Indeed, we have seen that the number of moves for insertion sort is always equal to n. This shows that it suffices to put c equal to any number greater than 1. □

Example 9.3. The number of data moves in selection sort is $O\{n\}$.

Proof. In fact, we have seen that the number of moves is $n-1$ at most. □

Example 9.4. The number of data moves in bubble sort is $O\{n^2\}$.

n	$n \log n$	n^2	n^{100}	2^n
1	0	1	1	2
10	10	100	10^{100}	1,024
100	200	10,000	10^{200}	1.6×10^{30}
1,000	3,000	1,000,000	10^{300}	1.1×10^{301}
10,000	40,000	100,000,000	10^{400}	$2.0 \times 10^{3,010}$
100,000	500,000	10,000,000,000	10^{500}	$10^{30,103}$

Figure 9.1. Growth of functions that could be used in order-of-magnitude statements.

Proof. We have seen above that the number of data moves is equal to at most

$$\frac{n(n-1)}{2} = \frac{n^2}{2} - \frac{n}{2}$$

We see that Definition 9.1 is satisfied with $f(n)$ equal to the number of data moves, $g(n)$ equal to n^2, and c equal to $1/2$. □

> *Q. Don't we lose too much information when we convert a precise count of data moves into an order-of-magnitude estimate?*
>
> *A. An order-of-magnitude estimate focuses attention on the most important aspect of running time—the behavior for large values of n. For example, the increase in running time caused by doubling n can be estimated merely from an order-of-magnitude estimate. With respect to data moves, if an algorithm—for example, insertion or selection sort— is $O\{n\}$, and we double n, then we tend to double the time spent in data moves. On the other hand, if—like bubble sort—the algorithm is $O\{n^2\}$ with respect to data moves, then doubling n causes the number of data moves to be quadrupled.*
>
> *For more complex algorithms, it may not be possible to give exact estimates of running time. In such cases, we must be content with order-of-magnitude estimates.*

Counting data comparisons

Data comparisons are more rapid than data moves in sorting playing cards by hand; however, data comparisons, because of their greater total number, could be more time consuming. In computer sorting, it is clearly necessary to count the number of data comparisons because there is not a great disparity in the time required for a data move as opposed to a data comparison.

1. For insertion sort, the greatest number of data comparisons occurs in case the initial order of the cards is from highest to lowest. In fact, each time we insert

a new card, we must compare it with *all* of the cards that have already been sorted. Inserting the first card requires no data comparisons, the second card requires one, the third card requires two, and so on. The total number of data comparisons in the worst case is $T_{n-1} = n(n-1)/2$. Thus, the running time of insertion sort with respect to data comparisons is $O\{n^2\}$—for sorting cards by hand and also for computer sorting. Strangely, it appears that the worst case is that the cards are initially in the correct order.

2. For selection sort, the worst case is that the cards are initially in the wrong order, from lowest to highest, because we search the unsorted cards from left to right—from highest to lowest. As above, the total number of data comparisons is no greater than $n(n-1)/2$. The worst case running time of selection sort with respect to data comparisons is $O\{n^2\}$.

3. For bubble sort, the worst case is that the cards are initially in the wrong order, from lowest to highest. As above, the number of data comparisons is $n(n-1)/2$, and the order of magnitude is $O\{n^2\}$.

Note that insertion, selection, and bubble sorts are all $O\{n^2\}$ with respect to data comparison.

Problem 9.5. In performing insertion sort by hand with 13 cards, estimate the total time spent (1) performing data moves (i.e., in physically moving the cards) and (2) performing data comparisons. Use the estimate of 0.5 moves per second and 0.01 data comparisons per second.

Solution.

1. Time for data moves $= n \times 0.5 = 13 \times 0.5 = 6.5$ seconds

2. Time for data comparisons $\leq \frac{1}{2}n(n-1) \times 0.01 = 78 \times 0.01 = 0.78$ seconds

This calculation confirms our expectation that, in sorting cards by hand, data moves take much more time than data comparisons. However, if the number of cards to be sorted is greatly increased, then the data comparisons will dominate.

For computer sorting applications we could lump together the number of data moves and the number of data comparisons because there is not a great time disparity between these two operations. For example, selection sort is $O\{n\}$ with respect to data moves and $O\{n^2\}$ with respect to data comparisons. What can be said about the order of magnitude of the sum? This question is answered by the following proposition.

Proposition 9.6. *Let f_1 and f_2 be nonnegative functions such that $f_1(n) = O\{n\}$ and $f_2(n) = O\{n^2\}$. Then $f_1(n) + f_2(n) = O\{n^2\}$.*

Proof. Since $f_1(n) = O\{n\}$, there exists a constant c_1 such that $f_1(n) < c_1 n$ for all sufficiently large values of n; similarly, there exists a constant c_2 such that $f_2(n) <$

$c_2 n^2$ for all sufficiently large n. Put $c = c_1 + c_2$; then, since $n \leq n^2$ for all natural numbers n, we have, for n sufficiently large,

$$f_1(n) + f_2(n) < c_1 n + c_2 n^2 \leq c_1 n^2 + c_2 n^2 \leq c n^2$$

which implies $f_1(n) + f_2(n) = O\{n^2\}$. □

More generally, in a *polynomial order of magnitude* such as $O\{4n^4 - 4n^2 + 5n + 6\}$, we can drop everything except the highest power of n. The order of magnitude $O\{n^4\}$ is equivalent. Any polynomial order of magnitude can be replaced by $O\{n^k\}$ where k is the highest exponent—the degree—of the polynomial. For certain problems of size n, there are algorithms for which running time has polynomial order of magnitude; sorting is a problem of this type. There are also problems for which there do not currently exist algorithms with polynomial running time. For such problems, we generally must be satisfied with algorithms with, for example, exponential running time. We are severely limited in our ability to solve problems of such large size.

There is a class of problems, called *NP hard*, for which currently there are no known polynomial algorithms. It is known that either all problems in this class, or none of them, have polynomial algorithms. However, it is generally thought that the second alternative holds—that none of them have polynomial algorithms. If a problem is NP hard, it is considered an indication of intractability. The following is an example of an NP-hard problem that is the subject of current research.

Example 9.7 (the traveling salesman problem). *NP hard.* A salesman must visit n different cities. Suppose that we know the cost of traveling from each city to every other city. Find a route for making a circuit of all n cities for the least total cost.

It follows from Proposition 9.6 that with respect to the sum of data moves and data comparisons, all the sorting algorithms we have considered so far—insertion, selection, and bubble sort—are $O\{n^2\}$. In the next section we will see that there are sorting algorithms that are $O\{n \log n\}$ with respect to the sum of moves and data comparisons. In this respect, these algorithms are better than the three sorting algorithms we have previously considered.

Divide-and-conquer algorithms

One of the most fruitful methodologies of science is to divide a complex problem into small parts that can be solved more easily and then to reintegrate those solutions into the larger problem. In computer science, this general methodology has a particular manifestation called *divide and conquer*, a strategy for the design of algorithms.

Card players tend to use a divide-and-conquer algorithm when they must sort a large number of cards. For example, to sort an entire deck of 52 cards, many players first sort the cards according to suit (spades, hearts, diamonds, clubs) and then according to rank.

Divide and conquer leads to several sorting methods, all of which are faster than insertion, selection, or bubble sort. We will discuss two methods known as mergesort and quicksort. We discuss mergesort first.

Mergesort

To mergesort a pile of items, we proceed as follows.

1. If there is only one item in the pile, then we are done because no sorting is needed.

2. *Divide* the pile into two subpiles. If there are an even number of items, divide the pile of items into two equal subpiles. If the number is odd, then put one more item in the second subpile than in the first.

3. *Sort* each pile separately with the largest items on top of each pile.

4. *Merge* the two piles as follows. The larger of the top two items in the subpiles is moved to become the top item in the merged pile. The larger of the remaining top two items of the subpiles becomes the second item of the merged pile, and so on.

This is *almost* a complete recipe for mergesort; however, one thing has been omitted. In step 3, we neglected to specify how to sort the subpiles. The answer is somewhat surprising. *We use mergesort to sort the subpiles.* At first glance, this seems to be unsatisfactory because we use mergesort as part of the definition of mergesort. A definition of this sort is called a *recursive definition*. Such a definition is permitted if the recursion always comes to an end after a finite number of steps. This is clearly the case with mergesort, because as we divide the piles we get smaller and smaller sized subpiles. Eventually, the size of the subpiles is one, and further division is impossible. Step 1 then ensures that the process ends after finitely many steps. Here is pseudocode for mergesort.

```
procedure Mergesort(pile)
    begin
        if pile has only one item
            then do nothing
        else
            Divide the pile into two subpiles
            Mergesort(first subpile)
            Mergesort(second subpile)
            Merge the two subpiles
    end
```

The computer is better suited than human beings for carrying out extensive recursive procedures. Our human brains soon become overwhelmed with keeping track of possibly hundreds of simultaneous mergesorts that are left hanging when further mergesorts are launched.

Current versions of higher level computer languages generally support recursion, but this has not always been true.[2] Every recursive procedure has an equivalent nonrecursive version. There can be technical reasons for preferring the nonrecursive version in certain cases, although the recursive version is usually easier to understand.

Figure 9.2 shows this process for the very small example in which we sort the letters in SORTING into alphabetical order, obtaining GINORTS. This problem is too small to demonstrate the speed advantage of mergesort.

Fact 9.8. *For a sorting problem of size n, the running time of mergesort, with respect to both data moves and data comparisons, is $O\{n \log n\}$.*

In the next section we consider *quicksort*, another divide-and-conquer algorithm with $O\{n \log n\}$ average running time.

Quicksort

The quicksort algorithm was invented by C. A. R. Hoare in 1962.[3] To quicksort a pile of items we proceed as follows.

1. If there is only one item in the pile, then we are done because no sorting is needed.

2. Choose an item at random from the pile. This item is called the *pivot*.

3. Form two subpiles as follows. On the immediate left of the pivot form a subpile containing all items from the original pile that have *higher* rank than the pivot. On the immediate right of the pivot form a subpile containing all items from the original pile of *lower* rank than the pivot. (It may happen that one of the subpiles is empty.)

1.			SOR-TING			
2.		S-OR	‖		TI-NG	
3.	S	O-R	‖	T-I		N-G
4.	S	O	R ‖ T	I	N	G
5.	S		OR ‖	IT		GN
6.		ORS	‖		GINT	
7.			GINORST			

Figure 9.2. Mergesort. This diagram shows seven steps in applying mergesort to the problem of rearranging the letters in SORTING into alphabetical order. In steps 1–4, the problem is divided into sub-, subsub-, and subsubsubproblems. In steps 5–7, first the subsubsub-, then the subsub-, and finally the subproblems are resolved using the merge procedure. Hyphens show the points at which divisions are made. The vertical double line shows the division of the main problem into two subproblems; the single vertical lines show the division into four subsubproblems; and additional spacing shows the further division into the seven subsubsubproblems consisting of single letters. Notice that the words ORS and GINT in line 6 are in alphabetical order because they are the results of merging the subsubproblems of line 5. In line 7, we merge ORS and GINT to obtain GINORST. In the merge process, we first compare the first letters of ORS and GINT. Since G comes before O, we move G to the first position of the result. We repeat the process with the remaining ORS and INT to find that the second letter of the result is I. And so on.

1.	QUICKSORT
2.	[I**C**] K {QUSO**R**T}
3.	C I K [Q**O**] R {ST**U**}
4.	CIKOQR[**S**T]U
5.	CIKOQRSTU

Figure 9.3. Quicksort. This diagram show five steps in applying quicksort to the problem of rearranging the letters in QUICKSORT in alphabetical order. In step 1, we choose, at random, **K** as a pivot point. In step 2, we place all letters higher than K in brackets to the left and all letters lower than K in braces to the right. At random, we choose two pivot points: **C** from the brackets, and **R** from the braces. In step 3, we place higher items to the left of the pivot and lower items to the right of the pivot within both the brackets and braces of step 2. This results in the interchange of I and C; the (nonpivot) letters in braces in step 2 are placed either in brackets or braces depending on whether they are higher or lower than the pivot R. Again in step 3, we choose the pivots at random, **O** and **U**. In step 4, pivoting about O interchanges Q and O, but pivoting about U leaves ST unchanged. We choose **S** as a pivot. In step 5, pivoting about S does nothing, because ST are already in the proper order.

4. Quicksort each of the two subpiles.

5. The items are now in the correct order from highest to lowest.

Quicksort has running time $O\{n\log n\}$ *on average.* In the unlikely event that the pivots are chosen badly, the running time could be $O\{n^2\}$, but this is not an important practical consideration.

Notice that quicksort, like mergesort, is a recursive algorithm because quicksort occurs in the definition of quicksort. Again the justification is that the successive subpiles become smaller and so must eventually consist of just one item. Quicksort is a divide-and-conquer algorithm because the original problem is replaced by smaller and smaller subproblems. Here is pseudocode for quicksort.

concatenate. To place side by side.

procedure Quicksort(pile)
    begin
        if pile has only one item
            then do nothing
        else
            Choose at random a pivot item from the pile
            Form a subpile of items higher than the pivot
            Form a subpile of items lower than the pivot
            Quicksort(first subpile)
            Quicksort(second subpile)
            Concatenate:
                First subpile on left.
                Pivot in the middle.
                Second subpile on right.
    end

Figure 9.3 shows an example of the quicksort algorithm applied to the problem of rearranging the letters in QUICKSORT in alphabetical order. Probably quicksort is too complex to be used in sorting playing cards by hand. However, when bridge players sort a hand of 13 cards, they usually follow a similar divide-and-conquer algorithm. First they sort the 13 cards according to suit and then each suit according to rank.

We have seen two examples of sorting algorithms that use the divide-and-conquer strategy.[4] There are divide-and-conquer algorithms for purposes other than sorting—for example, the Fast Fourier Transform (FFT).[5] The FFT algorithm is fundamental in digital signal processing (DSP), the basis of current digital sound and video technology.

In Figure 9.1 we see that computer science is sometimes concerned with very large numbers which are nevertheless finite. In the next chapter, we will see a world that contains not just one but an infinite multitude of *infinite numbers*.

Chapter 10

Set and Match

Man is equally incapable of seeing the nothingness from which he emerges and the infinity in which he is engulfed.

—BLAISE PASCAL, *Pensées* (1670)

Zeno's archery paradox initiated speculation and argument concerning the meaning of infinity, a recurrent riddle through the centuries for mathematicians, philosophers, and theologians—including the mathematician/philosopher/theologian, Blaise Pascal. In Chapter 2, we found that Zeno's speculations on the infinite fail under critical analysis, but it would be incorrect to suppose that infinity is beyond all scientific inquiry. The first rigorous mathematical theory of the infinite was discovered in the period 1871–84 by the German mathematician Georg Cantor, who devised a theory of not just one but infinitely many infinite numbers. The brilliant insight that made Cantor's theory possible is his beautiful *diagonalization method*. We will see applications of this method in this chapter.

In the previous chapters we have become familiar with different kinds of numbers. They fall into two classes:

1. Numbers that abstract the process of *counting*. First came the natural numbers, $1, 2, 3, \ldots$, and then the positive and negative integers.

2. Numbers that abstract the process of *measuring*. First came the rational numbers (the fractions) such as

$$-\frac{5}{13} \text{ or } \frac{41}{23}$$

and then came real numbers (the infinite decimal fractions). The real numbers are needed because rational numbers are not sufficient to measure the diagonal of a unit square.

We have not exhausted the mathematical generalizations that arise from counting. We use the natural numbers for counting 3 cows in a field, 7 pennies in a pocket, or generally the number of objects in a finite set, but what number can we use to describe the totality of all the natural numbers? Does it make sense to talk about infinite numbers? Yes, it does. Welcome to the surprising and sometimes controversial world of transfinite numbers.

Sets

Set is a mathematical term that means much the same as aggregate or collection. We indicate a set by putting the members (also called elements) of the set in curly brackets, for example, {Curly, Moe, Larry} and {Poppa bear, Momma bear, Baby bear}. Let's call these sets S (Stooges) and B (Bears).

How do we tell that the Stooges set has the same number of elements as the Bears set? The familiar process is to count the number of elements—three—in each set. This method fails with infinite sets because we cannot exhaust an infinite set by counting. However, there is another way of comparing the size of the two sets. We pair the elements of one set with the elements of the other set. Here is an example of such a pairing:

$$\text{Curly} \leftrightarrow \text{Poppa bear}$$
$$\text{Moe} \leftrightarrow \text{Momma bear}$$
$$\text{Larry} \leftrightarrow \text{Baby bear}$$

This pairing of elements of S with elements of B is an example of a *one-to-one* mapping of S onto B. Note that we have made an arbitrary choice of the direction of the mapping from S to B. This is a good place to make some definitions regarding mappings.[1]

Definition 10.1. Let f be a mapping from a set S into a set B.

1. If an element s of S is mapped by f to an element b of B, that is, if $b = f(s)$, then b is said to be the *image* of s.

2. If every element of B is the image of some element of S, then the mapping f is said to be from S *onto* B.

3. If no element of B is the image of more than one element of S, then the mapping f is said to be *one-to-one*.

4. If *every* element of S has an image in B, then the mapping f is said to be *defined everywhere* in S.

5. If f is *onto* and defined everywhere in S, then f is called a *one-to-one correspondence*.

It is often understood implicitly that a mapping is *defined everywhere*. Note carefully the usage of *into* and *onto*. If a mapping from S *into* B is not *onto* B it means that some element of B is not the image of any element of S. In mathematics, we observe the distinction between *into* and *onto* more carefully and consistently than in ordinary usage.

It is obvious that if there is a one-to-one correspondence between two *finite* sets, then the two sets have the same number of elements. If everyone at a party is dancing, and if every dancing couple consists of a boy and a girl, then, even if we have not counted them, we can be sure that there are exactly the same number of boys as girls. We apply this same idea to infinite sets. If there is a one-to-one correspondence between two infinite sets, then we are inclined to say that in some sense they have the same number of elements. More formally, we say that if there exists a one-to-one correspondence between two (finite or infinite) sets, then the two sets have the same *cardinality*. If S and T have the same cardinality, then we say that S is equivalent to T and we write

$$S \sim T$$

The relation $\sim$ shares certain properties with ordinary equality. These properties are called reflexivity, symmetry, and transitivity. We omit the proof of this very plausible theorem.

Theorem 10.2. *For any sets S, T, and R, the following statements are true.*

Reflexivity. *The set S is equivalent to itself; that is,*

$$S \sim S$$

Symmetry. *If S is equivalent to T, then T is equivalent to S; that is,*

$$\text{If } S \sim T, \text{ then } T \sim S$$

Transitivity. *If S is equivalent to T and T is equivalent to R, then S is equivalent to R; that is,*

$$\text{If } S \sim T \text{ and } T \sim R, \text{ then } S \sim R$$

The fact that these three properties are true ensures that all sets are divided into classes such that any two sets in the same class have the same cardinality. These classes are called *cardinal numbers*. For example, the cardinal number 3 *is* the class of all sets having three elements. The infinite set of natural numbers is associated with an infinite cardinal number that we will study in the following section.

Countability

The theory of cardinal numbers was created by Georg Cantor. He defined the standard notation, $\aleph_0$, for the cardinal number of the set of the natural numbers. He also discussed the cardinal number of the *real numbers*, the infinite decimals, which he denoted c. We will see that, in a certain sense, c is greater than $\aleph_0$.

The set of natural numbers is commonly denoted **N**.

Definition 10.3. A set that has a one-to-one correspondence with **N**, that is, a set with cardinal number $\aleph_0$, is said to be *countable*.

> **Notation.** Aleph $\aleph$ is the first letter of the Hebrew alphabet.
>
> The cardinal number of the set of the natural numbers, $\aleph_0$, is read "aleph-nought."
>
> The cardinality of the real numbers (the infinite decimals), also called the cardinality of the continuum, is denoted c.

To say that a set is countable means that one can conceive an infinite list of all the elements of the set. There should be a first, a second, and so on, and every element should be counted exactly once. This ordering in general has nothing to do with the usual numerical order. Such an order may not have any purpose other than to show that a set is countable.

For finite sets, it can never happen that a set has the same number of elements as one of its proper subsets. (A *proper subset* is a subset other than the empty set or the entire set. In other words, a proper subset of a set S is a nonempty subset of S that lacks at least one element of S.) Yet that is exactly what happens in the case of infinite sets. In fact, we can take this as a definition of an infinite set.

Definition 10.4. A set S is said to be *infinite* if there exists a one-to-one correspondence between S and a proper subset of S.

This definition is difficult because it does not seem to encompass our common-sense idea of *large size*. It is clear that no finite set satisfies Definition 10.4, but it is not immediately clear that this definition covers all sets that we generally consider infinite. To give a bit more credibility to Definition 10.4, we use it to show that **N**, the set of natural numbers, is an infinite set.

Proposition 10.5. *The set of natural numbers, **N**, is infinite.*

Proof. Define **N′** to be the set of all natural numbers except the number 1. Note that **N′** is a proper subset of **N**. Consider the one-to-one correspondence mapping between **N** and **N′** such that each natural number n in **N** corresponds to $n + 1$ in **N′**.

$$n \leftrightarrow n + 1$$

This mapping is one-to-one and onto. We have shown that the set of natural numbers, **N**, is infinite because we have shown that there exists a one-to-one correspondence of **N** onto a proper subset of **N**.   □

For **N′** and **N″** read "N prime" and "N double prime," respectively.

It might seem that in this proof one "last" number must be left over, but this does not happen because the set of natural numbers has no last number. Although the set of natural numbers, **N**, has one more element than **N′**, both sets have the same cardinal number $\aleph_0$.

To make this a little more vivid, suppose that there is an infinite hotel with all the natural numbers as room numbers. Suppose that there are no vacancies. Nevertheless, if you want a room, the room clerk can accommodate you. He just assigns room number 1 to you and for each natural number n he tells the guest in room n to move into room $n + 1$. Since there is no last natural number, every guest is accommodated.

It's even more surprising that the set of all even natural numbers, let us call it **N″**, has the same cardinality as the set **N**. This can be seen by the following one-to-one correspondence:

$$n \leftrightarrow 2n$$

Q. *Common sense says that **N** is* twice *the size of **N″**. Why do cardinal numbers deny this obvious fact?*

A. The concept of cardinal number is a high level abstraction that blurs this distinction.

Q. Are cardinal numbers so abstract that all infinite sets are lumped together? Do all infinite sets have the same cardinal number?

A. No. On the contrary, we will see that there are infinitely many different cardinal numbers.

There are many sets (e.g., N', N'') that have a one-to-one correspondence with N, and all these sets have cardinality $\aleph_0$. Since N' are N'' are proper subsets of N, we see that, paradoxically, it is possible for a set to have the same cardinality as its proper subset—in contradiction to our expectation that "the whole is greater than its parts."

The previous examples are fairly plausible, but the following example is very surprising. There are reasons to expect that the number of rational numbers is greater than the number of integers. In fact, not only are the integers a proper subset of the rationals, but it is also true that between any two integers there are infinitely many rational numbers. Nevertheless, using Cantor's diagonalization method we are able to prove that both sets have the same cardinal number $\aleph_0$. This assertion is equivalent to the following.

rational numbers. Recall that the rational numbers are the numbers of the form p/q, where p and q are integers.

Theorem 10.6. *The set of rational numbers is countable.*

It is simpler to prove the following seemingly weaker proposition.

Proposition 10.7. *The set of positive rational numbers is countable.*

Proof. Let us denote the set of positive rational numbers Q_+. (The letter Q stands for quotient—the rational numbers are quotients of natural numbers. The letter R is reserved for the real numbers.) The set Q_+ consists precisely of the numbers that can be expressed in the form p/q, where p and q are natural numbers. For each natural number n greater than 1, let us define a subset Q_n of Q_+—that is, a subset of the set of positive rational numbers—as follows. For each natural number n greater than 1, define Q_n to consist of all rational numbers that can be expressed as p/q such that

$$p+q=n$$

In other words, a fraction belongs to Q_n if the sum of its numerator and denominator is n. The reason for excluding $n = 1$ is clear. If we defined Q_1, it must be the empty set because there are no natural numbers p and q such that $p + q = 1$. (Recall that the smallest natural number is 1.)

Figure 10.1 shows lists of the elements of each of the subsets Q_n, arranged according to increasing values of n. There are several things to notice about these subsets $Q_n(n = 2, 3, \ldots)$ of the positive rational numbers:

- Each positive rational number belongs to at least one of the subsets Q_n. Some positive rational numbers occur in more than one subset Q_n because some of the fractions listed in Figure 10.1 are not in lowest terms; that is, in some cases the numerator and denominator have a common factor. If the fractions not in lowest terms were removed from Figure 10.1, then each positive rational number would occur *exactly* once.

$$
\begin{array}{lllllll}
\mathbf{Q_2}: & 1/1 \\
\mathbf{Q_3}: & 1/2 & 2/1 \\
\mathbf{Q_4}: & 1/3 & 2/2 & 3/1 \\
\mathbf{Q_5}: & 1/4 & 2/3 & 3/2 & 4/1 \\
\mathbf{Q_6}: & 1/5 & 2/4 & 3/3 & 4/2 & 5/1 \\
\mathbf{Q_7}: & 1/6 & 2/5 & 3/4 & 4/3 & 5/2 & 6/1 \\
& \cdot & \cdot & \cdot & \cdot & \cdot & \cdot
\end{array}
$$

Figure 10.1. Counting the positive rational numbers.

- Each subset $\mathbf{Q}_n$, that is, each row of Figure 10.1, contains only finitely many positive rational numbers. This is true because for each natural number n there are exactly $n-1$ pairs of natural numbers (p,q) such that $p+q=n$. For example, for $n=5$, there are the pairs

$$(1,4),(2,3),(3,2), \text{ and } (4,1)$$

We show a one-to-one correspondence between the natural numbers and the positive rational numbers as follows. We define the one-to-one correspondence by counting row by row the numbers in Figure 10.1 from left to right, skipping any of the fractions that are not in lowest terms. The fractions in lowest terms are shown in Figure 10.2 in boxes; the corresponding natural numbers are shown as subscripts to the boxes. Every rational number eventually appears in this array. We ensure that it is counted just once by skipping the fractions not in lowest terms.

$$
\begin{array}{lllllll}
\mathbf{Q_2}: & \boxed{1/1}_1 \\
\mathbf{Q_3}: & \boxed{1/2}_2 & \boxed{2/1}_3 \\
\mathbf{Q_4}: & \boxed{1/3}_4 & 2/2 & \boxed{3/1}_5 \\
\mathbf{Q_5}: & \boxed{1/4}_6 & \boxed{2/3}_7 & \boxed{3/2}_8 & \boxed{4/1}_9 \\
\mathbf{Q_6}: & \boxed{1/5}_{10} & 2/4 & 3/3 & 4/2 & \boxed{5/1}_{11} \\
\mathbf{Q_7}: & \boxed{1/6}_{12} & \boxed{2/5}_{13} & \boxed{3/4}_{14} & \boxed{4/3}_{15} & \boxed{5/2}_{16} & \boxed{6/1}_{17} \\
& \cdot & \cdot & \cdot & \cdot & \cdot & \cdot
\end{array}
$$

Figure 10.2. The boxes enumerate all the positive rational numbers with no repetitions.

This concludes the proof because we have shown a one-to-one correspondence between the positive rational numbers and the natural numbers. $\square$

The above result is unexpected since there appear to be vastly more rational numbers than natural numbers because between any two natural numbers there are infinitely many rational numbers. It is natural now to ask whether every infinite set is countable. Are there infinite sets that are *not* countable? The answer is Yes. In fact, it was stated earlier that the cardinal number c of the set of all real numbers is *greater* than $\aleph_0$.

Let us define what it means to say that one cardinal number is greater than another cardinal number.

Definition 10.8. Let S and T be sets. We say that the cardinal number of S is less than the cardinal number of T if the following two conditions hold.

1. There exists a one-to-one correspondence between S and some subset of T.

2. There does not exist a one-to-one correspondence between S and the entire set T.

In particular, to show that c is greater than $\aleph_0$, it is sufficient to show the following.

1. Show that there is a one-to-one correspondence between the natural numbers and a subset of the real numbers. But this is very easy because the natural numbers are a subset of the real numbers. The required correspondence is simply the following:

$$1 \leftrightarrow 1.000\ldots$$
$$2 \leftrightarrow 2.000\ldots$$
$$3 \leftrightarrow 3.000\ldots$$
$$\cdot \quad \cdot \quad \cdot$$

In other cases that we will consider it is equally easy to find a one-to-one correspondence between the smaller set and a subset of the larger set.

2. Show that there cannot be a one-to-one correspondence between the natural numbers and the real numbers. This is not at all obvious, and it is not very easy to show.

The following result from Cantor is quite surprising since it is difficult to find a reason to believe that the real numbers are more numerous than the rational numbers because between any two rational numbers there are infinitely many real numbers (infinite decimals), and between any two real numbers there are infinitely many rational numbers.

Let's state Cantor's result more formally.

Theorem 10.9. *There does not exist a one-to-one correspondence between the set of natural numbers and the set of real numbers.*

Proof. The proof again uses Cantor's beautiful *diagonalization method.* The proof is indirect; that is, we suppose that the result is false and we derive a contradiction. Suppose that there were such a correspondence. This means that there must be an infinite list

$$R_1, R_2, R_3, \ldots$$

that contains every real number (infinite decimal) exactly once. Since the real numbers are identical to the infinite decimals, suppose that a particular correspondence is as shown in Figure 10.3.

Integral part	Fractional part

$$
\begin{aligned}
1 &\leftrightarrow R_1 = N_1.\mathbf{a_1}\, a_2\, a_3\, a_4\, a_5\, a_6\, a_7 \ldots \\
2 &\leftrightarrow R_2 = N_2.b_1\, \mathbf{b_2}\, b_3\, b_4\, b_5\, b_6\, b_7 \ldots \\
3 &\leftrightarrow R_3 = N_3.c_1\, c_2\, \mathbf{c_3}\, c_4\, c_5\, c_6\, c_7 \ldots \\
4 &\leftrightarrow R_4 = N_4.d_1\, d_2\, d_3\, \mathbf{d_4}\, d_5\, d_6\, d_7 \ldots \\
5 &\leftrightarrow R_5 = N_5.e_1\, e_2\, e_3\, e_4\, \mathbf{e_5}\, e_6\, e_7 \ldots \\
6 &\leftrightarrow R_6 = N_6.f_1\, f_2\, f_3\, f_4\, f_5\, \mathbf{f_6}\, f_7 \ldots
\end{aligned}
$$

$$\cdot \quad \cdot \quad \cdot$$

Figure 10.3. Cantor's diagonalization method shows that the real numbers are uncountable.

The *integral part* of a real number x is defined to be the largest whole number not larger than x. The *fractional part* of x is defined to be x minus the integral part of x. For example, the integral part of $\pi = 3.14159\ldots$ is 3 and its fractional part is $0.14159\ldots$. In Figure 10.3, the integral parts of the real numbers

$$R_1, R_2, R_3, \ldots$$

are

$$N_1, N_2, N_3, \ldots$$

respectively, and the digits of the fractional parts are

$$a_1, a_2, a_3 \ldots$$
$$b_1, b_2, b_3 \ldots$$
$$c_1, c_2, c_3 \ldots$$

$$\cdot \quad \cdot \quad \cdot$$

We construct a real number R that is not in the list by the following method—Cantor's diagonalization method. We define R by defining separately its integral part and its fractional part. The integral part of R is unimportant; define it to be zero. We define the fractional part by specifying the digits of the decimal expansion of R:

$$R = 0.abcdef\ldots \tag{10.1}$$

Notation. $\pi = 3.14159\ldots$.
Integral part:
$[\pi] = 3$.
Fractional part:
$\{\pi\} = 0.14159\ldots$.

The digits to the right of the decimal point, $a, b, c, \ldots$, are chosen by looking at the diagonal digits in the list, the digits that are shown in Figure 10.3 in boldface. The idea is to define R so that

R differs from R_1 in the 1st decimal place,
R differs from R_2 in the 2nd decimal place,
R differs from R_3 in the 3rd decimal place,   (10.2)

. . .

R differs from R_n in the nth decimal place,

. . .

We construct the digits of R using only 0's and 1's, but we could use any pair of the ten possible digits except that we wish to avoid using the digit 9 so that we don't have to be concerned with the possibility of dual representations of numbers—for example,

$$0.9999\ldots = 1.0000\ldots$$

We choose the digits as follows. Note that $a, b, c, \ldots$ refer to equation (10.1) and $a_1, b_2, c_3, \ldots$ refer to Figure 10.3.

Put $a = 1$ if $a_1 = 0$, otherwise put $a = 0$.
Put $b = 1$ if $b_2 = 0$, otherwise put $b = 0$.
Put $c = 1$ if $c_3 = 0$, otherwise put $c = 0$.
Put $d = 1$ if $d_4 = 0$, otherwise put $d = 0$.
Put $e = 1$ if $e_5 = 0$, otherwise put $e = 0$.
Put $f = 1$ if $f_6 = 0$, otherwise put $f = 0$.

. . .

The real number R constructed in this way satisfies the conditions (10.2). It follows that R is not any one of the numbers on the list of Figure 10.3. Thus, the existence of this list, which is equivalent to the assertion that Theorem 10.9 is false, has led us to a contradiction, and we are finished with the proof.   □   ↩ Dangerous curve!

The foregoing argument can be modified to show that Cantor's middle-thirds set is also uncountable.[2]

It is natural now to ask if there is any set that has greater cardinality than the set of real numbers. In the following section we will answer this question.

The Set of All Subsets

> **Notation**. If S is a set, then 2^S is the set of all subsets of S.
>
> ---
>
> Recall that the empty set is denoted $\emptyset$.

The answer is Yes, there are sets that are larger than the set of the real numbers. In particular, the set of all subsets of the real numbers is larger than the set of real numbers. Indeed, we will show that for any nonempty set S, there cannot be a one-to-one correspondence between S and the set of all the subsets of S. The set of all of the subsets of S is denoted 2^S. Of course, this expression is not an exponential in the usual sense. We simply use 2^S as a notation for the set of all subsets of S. We can see that this notation is reasonable because in Proposition 8.4 we see that the number of subsets of a set containing n elements is 2^n. In fact, let us review the case $n = 3$. Let S be the set $\{1, 2, 3\}$. Here is a list of all of the subsets of S:

$$\emptyset, \ \{1\}, \ \{2\}, \ \{3\}, \ \{1, 2\}, \ \{1, 3\}, \ \{2, 3\}, \ \{1, 2, 3\}$$

These are the elements of the set 2^S, that is, the set of all subsets of S. Notice that the number of elements in S is 3 and the number of elements in 2^S is equal to 2^3.

For the finite set with three elements just discussed, it is clear that there are more members in the set of all subsets than there are in the set itself, and this is obviously true for all finite sets. The case of infinite sets is much more interesting.

For infinite sets, we have seen examples in which a set can have the same cardinality as a set that appears to be much smaller. For example, the set of rational numbers has the same cardinality as the set of natural numbers. Nevertheless, the following theorem is true.

Theorem 10.10. *For any nonempty set S, there cannot be a one-to-one correspondence between S and the set of all subsets of S.*

Theorem 10.10 doesn't make any distinction between finite and infinite sets. Since the finite case is easy, the interest of this theorem lies in the fact that it is true for *infinite* as well as finite sets.

The monopolistic barber

The proof of Theorem 10.10 is tricky. It may help to interject here a little story that relates to a famous paradox of Bertrand Russell (1872–1970).

The Russell paradox brought about a complete breakdown and reformulation of set theory as it was known in the first decade of this century. We won't discuss this matter systematically, but here is a story that illustrates the Russell paradox and prepares for the proof of Theorem 10.10.

Example 10.11. In a certain town there is a barber who shaves all men of the town who do not shave themselves. *Prove that the barber must be a woman.*

Proof. We proceed by the indirect method. Suppose that, on the contrary, the barber is a man.

1. If he shaves himself, then he doesn't shave himself, because he only shaves those men who do not shave themselves. Contradiction.

2. On the other hand, if he doesn't shave himself, then he does shave himself, because he shaves all men who do not shave themselves. Contradiction again.

Since the assumption that the barber is a man leads to a contradiction, it follows that the barber is a woman.   □

The planet of the dogs

Here is another little story in a similar vein. Zargon the astronaut has returned from a journey to the planet Pluto. He makes a number of claims about the planet, but we do not believe him. In fact, we show that his claims are false because they contain a subtle contradiction. Here are Zargon's claims about the planet Pluto:

- Pluto is the home of infinitely many dogs. Some of the dogs are brown and some of them are white.

- There are very rigid rules on Pluto that prescribe when a dog can sniff another dog. Each dog is only permitted to sniff dogs on his sniff-list.

- No two sniff-lists are the same.

- For each subset of the set of dogs on Pluto, there is a dog whose sniff-list is precisely that subset.

- Some dogs are on their own sniff-lists. They are permitted to sniff themselves, and they are colored brown.

- All other dogs are not on their own sniff-lists. They are not permitted to sniff themselves, and they are colored white.

How can we prove that Zargon is a liar? His assertion contains a contradiction as follows.

The set of all the white dogs must be the sniff-list for some dog; let's call him Fido. *What is the color of Fido? Is he white or is he brown?*

If Fido is brown, it means that Fido is on his own sniff-list. But this is impossible because all the dogs on his sniff-list are white.

If Fido is white, it means that Fido is not on his own sniff-list. But this is impossible because all the dogs not on his sniff-list are brown.

There are no other possibilities, so the planet of the dogs described by Zargon is impossible.

We will soon see the relevance of this story. Let's return to the proof of Theorem 10.10. Now would be a good time to reread the statement of Theorem 10.10.

Proof of Theorem 10.10. Let us suppose that this theorem is false. In other words, suppose that there is such a one-to-one correspondence between the set S and the set ↩ Dangerous curve! of all subsets of S. We will show that this assumption leads to a contradiction and that, therefore, the theorem must be true. Under the assumed one-to-one correspondence, let T be a subset of S consisting of all elements of S that do *not* belong to the subset to which they correspond. (Think of the subset T as the set of white dogs on Pluto.) There must be an element t of S that corresponds to T under the assumed one-to-one correspondence. (Fido is the element t.)

Either t belongs to T or not, but we will see that either possibility leads to a contradiction.

- If t belongs to $\mathcal{T}$, then t fails the condition that defines membership in $\mathcal{T}$. Contradiction!

- If t does not belong to $\mathcal{T}$, then t meets the condition that defines membership in $\mathcal{T}$. Contradiction again! □

Theorem 10.10 has a number of consequences.

- It can be shown that the set of all subsets of the natural numbers has the same cardinality as the set of real numbers. Having shown this, Theorem 10.10 implies Theorem 10.9.

- The cardinality of a nonempty set is less than the cardinality of the set of all its subsets. We must show that the two conditions of Definition 10.8 are satisfied. Condition 1 is easy to show, but we will not give a formal proof. Condition 2 of Definition 10.8 is true because of Theorem 10.10.

- Theorem 10.10 answers the question, "Are there sets that are larger than the set of real numbers?" Theorem 10.10, together with the paragraph above, shows, as claimed earlier, that the set of all subsets of the real numbers is a larger set than the set of real numbers.

In this chapter, we have used cardinal numbers to classify sets, finite and infinite, according to size. We have seen several surprising results that follow from Cantor's diagonalization method. In the next chapter, we will see that set theory also provides a basis for understanding games of chance.

Chapter 11

Chance Encounter

Life is the art of drawing sufficient conclusions from insufficient premises.
—SAMUEL BUTLER (1835–1902), *Note-Books*

In the sixteenth century, Girolamo Cardano (1501–76) was the first to write on the laws of chance.[1] However, the modern theory of probability started in 1654, when a French nobleman, the Chevalier de Méré, addressed questions concerning a dice game to the mathematician Blaise Pascal.[2] Pascal engaged in a correspondence with Fermat about the matter. One of the questions that de Méré asked is this.

Problem 11.1. Which of the following is more likely?

1. In one throw of four dice to get at least one six.

2. In twenty-four throws of two dice to get at least one double-six.

The Chevalier thought that the two were equally likely, a mistake that cost him money at the gambling table. We will answer his question completely, but first we will discuss the *fundamental concepts of probability.*

Sample Space and Probability

The universe of all possible outcomes of an experiment is called the *sample space* for that experiment. For example, we define the sample space of the possible outcomes of throwing two dice, a red die and a blue die. The sample space S_1, shown in Figure 11.1, consists of 36 ordered pairs showing the number on the red die first and the number on the blue die second. Note that the pairs $(1, 2)$ and $(2, 1)$ are represented by two different points of the sample space. The important reason for keeping these pairs distinct is to ensure that the points in this sample space represent equally likely events.

Sample spaces can be very complex, but this one is simple for two reasons.

1. It contains finitely many points.

> In mathematics there are many abstractions that are called *spaces*: sample space, measure space, vector space, Hilbert space, Banach space, and so on. These concepts are far-reaching generalizations of the three-dimensional space in which we live.

$$
\begin{array}{cccccc}
(1,1) & (1,2) & (1,3) & (1,4) & (1,5) & (1,6) \\
(2,1) & (2,2) & (2,3) & (2,4) & (2,5) & (2,6) \\
(3,1) & (3,2) & (3,3) & (3,4) & (3,5) & (3,6) \\
(4,1) & (4,2) & (4,3) & (4,4) & (4,5) & (4,6) \\
(5,1) & (5,2) & (5,3) & (5,4) & (5,5) & (5,6) \\
(6,1) & (6,2) & (6,3) & (6,4) & (6,5) & (6,6)
\end{array}
$$

Figure 11.1. The sample space S_1 for one throw of two dice.

2. The points represent *equally likely* outcomes of the experiment. This is a mathematical model that is abstracted from the actual physical dice. We ignore the possibility that, either by accident or design, the dice have irregularities such that these outcomes are not equally likely.

pip. Any of the spots on dice, dominos, or playing cards.

A subset of the sample space is called an *event*. For example, the event in which the pips on two dice total six is the subset A_1 containing the five pairs

$$(5,1), (4,2), (3,3), (2,4), (1,5)$$

In this technical sense, the word event, *like subset, is a singular collective noun, even when it refers to a multiplicity of points. An event that refers to a single point in the sample space is called an* atomic *event.*

We will confine our attention to sample spaces like S_1 in which the points are equally likely.

Definition 11.2. The number of points in a finite set A is denoted $N[A]$.

Definition 11.3. Let S be a finite nonempty sample space in which the single points are equally likely events, and let A be an arbitrary event of S. We define the *probability* of A, which we denote $P[A]$, as follows.

$$P[A] = \frac{N[A]}{N[S]} \tag{11.1}$$

Loosely speaking, the probability of an event is a fraction that tends to match the fraction of trials in which the event occurs, that is, the number of successful trials divided by the total number of trials. The word *trials* here is just a useful fiction for visualizing probability. We define probabilities of events even if there may never be even one trial in the ordinary sense.

Example 11.4. On throwing two dice, find the probabilities of each of the following two events.

A_1: The total of the two dice is six.

B_1: The dice show neither double-five nor double-six.

Solution. The sample space is S_1 in Figure 11.1. The number $N[S_1]$ of points in S_1 is 36.

A_1: As seen above, the number $N[A_1]$ of points in the event A_1 is 5. Therefore the probability $P[A_1]$ is equal to

$$\frac{N[A_1]}{N[S_1]} = \frac{5}{36}$$

B_1: The set B_1 is shown in Figure 11.2. The number $N[B_1]$ is equal to 34. Therefore, the probability $P[B_1]$ is equal to $34/36$.

$$
\begin{array}{llllll}
(1,1) & (1,2) & (1,3) & (1,4) & (1,5) & (1,6) \\
(2,1) & (2,2) & (2,3) & (2,4) & (2,5) & (2,6) \\
(3,1) & (3,2) & (3,3) & (3,4) & (3,5) & (3,6) \\
(4,1) & (4,2) & (4,3) & (4,4) & (4,5) & (4,6) \\
(5,1) & (5,2) & (5,3) & (5,4) &   & (5,6) \\
(6,1) & (6,2) & (6,3) & (6,4) & (6,5) &
\end{array}
$$

Figure 11.2. The sample space B_1 for Examples 11.4 and 11.7.

We have defined probability for events in a finite sample space *with all points equally likely.* A sample space together with a probability function is called a *probability space.* In a more abstract development of probability theory, a probability space is a sample space, finite or infinite, together with a probability function that satisfies certain axioms and that is defined on certain subsets of the sample space. We will not elaborate further on this more abstract setting of probability theory, but we will use the term probability space because we want to emphasize the existence of a probability function.

The concrete examples in this chapter all deal with finite probability spaces in which the probability function is given by Definition 11.3.

The Algebra of Sets

We have seen that probability theory is concerned with certain sets called sample spaces. Certain concepts of set theory are necessary before we can proceed. We wish to define certain modes of combining given sets (events) to form new sets. In particular, we introduce the relation $\subset$ and the operators $\cap$ and $\cup$. The subsets of the sample space S together with these operators define an *algebra of sets* or a Boolean algebra after George Boole (1815–64) who created this concept.

Definition 11.5. Let S be a set, and let A and B be subsets of S. (One can equally well say, Let S be a sample space and let A and B be events; S is also called the *certain event* because a sample space represents everything that can possibly happen.)

1. If A is a subset of B then we write $A \subset B$.

2. The intersection of A and B, written $A \cap B$, is defined as the set of points of S that belong to both A and B.

3. The union of A and B, written $A \cup B$, is defined as the set of points of S that belong to either A or B.

4. The difference A minus B, written $A \sim B$, is defined as the set of points of S that belong to A but not to B.

5. The set $S \sim A$, that is, the set of points in S that do not belong to A, is called the complement of A and is denoted A^c.

6. The empty set is denoted $\emptyset$.

Probability and Conditional Probability

The probability function P has a number of properties that are easy to derive from Definition 11.3.

Fact 11.6. *Let S be a finite probability space with P defined by equation* (11.1). *For every subset A and B of S the following properties hold.*

1. *$P[A] \geq 0$. That is, probability is never negative.*

2. *$P[S] = 1$. That is, the probability of the certain event is 1.*

3. *If $A \cap B = \emptyset$, then $P[A \cup B] = P[A] + P[B]$. That is, if A and B are disjoint sets (mutually exclusive events), then the probability of the union is equal to the sum of the probabilities of A and B.*

For a more general probability space, for example, a space that is infinite or for which the points are not equally likely, the items of Fact 11.6 are taken as axioms for the probability function.

Conditional probability

The following modification of Example 11.4 leads us to the concept of conditional probability. Since this problem is quite simple, it is easy to solve either with or without conditional probability.

Example 11.7. Throw two dice until the result is *not* a double-six or a double-five. What is the probability that on the final throw the sum of the pips is six?

Solution. The sample space B_1 for this problem is shown in Figure 11.2—Figure 11.1 without the two pairs $(6, 6)$ and $(5, 5)$. The number of points in A_1 is five, the same as in Example 11.4, and the number of points in the sample space B_1 is 34. Thus, according to Definition 11.3, the desired probability is $5/34$.

Examples 11.4 and 11.7 are related. This relationship is obscured if we define B_1 as a completely new sample space. The following concept enables us to solve the problem without defining a new sample space.

Definition 11.8. Let S be a probability space with probability function P, and let A and B be events such that $P[B] > 0$. The conditional probability of A given B, which we write $P[A \mid B]$, is defined as follows.

$$P[A \mid B] = \frac{P[A \cap B]}{P[B]}$$

A different statement of the result of Example 11.7 is that the conditional probability that the sum of the pips on two dice is six, given that it is not a double-six or a double-five, is 5/34. To verify this, we use the above definition with the events $A = A_1$ and $B = B_1$ from Example 11.4. In that example we saw that the probability $P[A_1]$ is equal to 5/36 and the probability $P[B_1]$ is equal to 34/36. Since $A_1 \subset B_1$, we have $A_1 \cap B_1 = A_1$. Now we use Definition 11.8 to compute the conditional probability

$$P[A_1 \mid B_1] = \frac{P[A_1 \cap B_1]}{P[B_1]} = \frac{5/36}{34/36} = \frac{5}{34}$$

Example 11.9. Find the probability of throwing a double-six with two dice colored red and blue:

1. Given that the red die is a six.

2. Given that at least one of the dice is a six.

Solution of 1. In Figure 11.1, the subset B_1 for which the red die is a six consists of the six points in the last row of S_1, and the subset A for which both dice are six consists of the single point $(6,6)$. Thus, we have

$$P[A \mid B_1] = \frac{P[A \cap B_1]}{P[B_1]} = \frac{N[A]}{N[B_1]} = \frac{1}{6}$$

Solution of 2. In Figure 11.1, the subset B_2 for which at least one die is a six consists of the eleven points in the last row together with the last column of S_1. The subset A for which both dice are six again consists of the single point $(6,6)$. Thus, we have

$$P[A \mid B_2] = \frac{P[A \cap B_2]}{P[B_2]} = \frac{N[A]}{N[B_2]} = \frac{1}{11}$$

The Monty Hall Problem

Many find the solution of Example 11.9 puzzling. It seems that we do not possess a natural intuition for conditional probability. In fact, the uncritical application of conditional probability can easily lead to errors. That is the key to a probability puzzle[3] that has created a stir recently in the popular press and misled many including a few professional mathematicians. This puzzle, called the *Monty Hall problem*, is the following.

Problem 11.10. A television game show host asks a contestant to choose one of three closed doors to obtain the prize concealed therein. Behind two of the doors are goats and behind one of them is a new automobile. The contestant chooses one of the doors, but then, before she opens that door, the host shows the contestant a goat behind one of the other two doors. The host then asks the contestant if she would like to switch her choice to the remaining door. Should she switch? What is her probability of winning the car if she switches? What is it if she doesn't switch?

Solution. The problem as it stands does not have quite enough information for a solution. The host, who knows where the car and the goats are, can influence the outcome by waiting to see the contestant's choice and only then deciding whether to offer the option to switch.

We wish to discuss the version of the problem in which the host promises in advance that he will in any case offer the option to switch. In case the contestant initially chooses the car, the host flips a coin to determine which of the two goats to show—heads he opens the larger numbered door, and tails he opens the smaller numbered door.

The contestant frequently believes the following.

1. Before she is shown the goat, she believes that her probability of choosing the car is $1/3$.

2. After she is shown the goat, she believes that her probability of choosing the car is $1/2$ no matter whether she switches or not. She arrives at this conclusion because she believes that it is equally likely that the car is behind either of the two remaining doors.

In fact, 1 is correct, but 2 is incorrect. We will see that the contestant's probability of obtaining the car is $2/3$ if she switches and $1/3$ if she does not switch.

The correct solution of the problem is very simple. The sample space consists of three points. Let's call the goats Arlo and Bertha. Then we will call the three equally likely points in the sample space:

$$S_1 = \{\text{Arlo, Bertha, Car}\}$$

The contestant can choose in advance which strategy to use, switch or no-switch, because no additional information is gained from seeing the goat. Let us examine the payoff of each of these strategies

1. Suppose the contestant decides to switch. Then she wins the car if her initial choice is a door that conceals a goat. The probability of making this choice is $2/3$.

2. Suppose the contestant decides *not* to switch. Then she wins the car if her initial choice is the car. The probability of making this choice is $1/3$.

For one who wishes to confirm the above solution using conditional probability, the following hint may be helpful. This problem is tricky because it is easy to misuse the concept of conditional probability by not accounting properly for the case in which the host must toss a coin to decide which goat's door to open.

Before we look at more probability problems, we should discuss *independence*, a concept that will help us to solve Problem 11.1, the problem that de Méré posed to Pascal.

Independence

The sample space S_1 in Figure 11.1 for the outcomes of throwing two dice illustrates a very important concept in probability theory called *independence* which we will define shortly (Definition 11.11). The third row of S_1 represents the event R_3 that the red die shows a three, and the fourth column of S_1 represents the event C_4 in which the blue die shows a four. Common sense says that these two events "have nothing to do with each other." This is reflected in the following relationships involving probabilities. Noting that $R_3 \cap C_4$ contains just the one point $(3,4)$, we compute the conditional probability that the blue die is a four, given that the red die is a three, as follows.

$$P[C_4 \mid R_3] = \frac{P[R_3 \cap C_4]}{P[R_3]} = \frac{1}{6}$$

Since $P[C_4] = 1/6$, we see that $P[C_4 \mid R_3] = P[C_4]$; that is, the conditional probability is the same as the unconditional probability. It follows that

$$P[R_3 \cap C_4] = P[R_3]P[C_4]$$

We use this relationship as a basis for the definition of independence.

Definition 11.11. Let S be a probability space with probability function P, and let A and B be events. We say that A and B are *independent* if

$$P[A \cap B] = P[A]P[B]$$

Fact 11.12. *If $P[B]$ is nonzero, then A and B are independent events if and only if*

$$P[A \mid B] = P[A]$$

Proposition 11.13. *Mutually exclusive events with nonzero probabilities are never independent.*

Proof. To say that A and B are mutually exclusive means that they have an empty intersection, that is, $A \cap B = \emptyset$, which implies $P[A \cap B] = 0$. Hence we must have $P[A \mid B] = 0$. But, by Fact 11.12, it follows that $P[A] = 0$, contrary to our assumption that the probabilities are nonzero.   □

> *n*-tuple. An ordered list
> $$(a_1, \ a_2, \ldots, a_n)$$
> of *n* numbers or other mathematical objects all belonging to the same underlying set.

The outcome of throwing a single die n times is expressed as an *n*-tuple

$$(a_1, a_2, \ldots, a_n)$$

whose components are integers between 1 and 6. The normal assumption is that the probability of throwing a particular number on one throw is independent of the probability of throwing another (or the same) particular number on another throw. We express this fact by saying that this experiment consists of *n independent trials*.

Definition 11.14. The sample space S_n of *n*-tuples from a set S is called a sample space of *n independent trials* if for arbitrary x, y in S and arbitrary natural numbers i, j not greater than n, the events $s_i = x$ and $s_j = y$ are independent.

De Méré's Problems

The problems of de Méré (Problem 11.1) are most easily solved using the concept of independence. We will solve the first problem first without and then with the use of independence.

Solution of Problem 11.1.1. We wish to find the probability of throwing at least one six with four dice. The sample space S for this problem consists of all possible throws of four dice. It is useful to distinguish the individual dice by giving them distinctive colors: red, blue, green, yellow. Thus, there are separate points in the sample space for the throws $(r1, b2, g3, y4)$ and $(r4, b3, g2, y1)$, events that would be indistinguishable on the gambling table if the dice were all of one color. The sample space has too many points in it to show conveniently in a diagram, but, nevertheless, we can count them. The red die can show six different numbers, and, for each of these, the blue die can show six different numbers, and so on. Altogether, there are $6^4 = 1296$ points in the sample space.

Now consider the subset A of S that contains at least one six. It is easier to count the complement A^c of this subset, the set of throws with no six. By the same method that we used above for counting the entire set, we find that the event A^c—throws of four dice with no six—corresponds to $5^4 = 625$ points in the sample space. It follows that the set A of throws with at least one six has $1296 - 625 = 671$ points in the sample space. The probability of throwing at least one six is equal to

Notation. The symbol $\approx$ is read "is approximately equal to." A formula that includes a decimal fraction on the right of $\approx$ also asserts the accuracy of the approximation within the normal roundoff error. For example, when we write

$$\pi \approx 3.1416$$

we are asserting that π is less than 3.14165 but not less than 3.14155; in other words,

$$3.14155 \leq \pi < 3.14165$$

$$P[A] = \frac{N[A]}{N[S]} = \frac{671}{1296} \approx 0.51775$$

We arrive at the same result by considering independent trials. The probability of not throwing a six with one die is $5/6$. From independence, the probability of not throwing six in four trials is

$$P[A^c] = \left(\frac{5}{6}\right)^4 \approx 0.48225$$

Consequently the required probability is

$$P[A] = 1 - P[A^c] \approx 0.51775$$

Solution of Problem 11.1.2. We want to find the probability of at least one double-six in 24 throws of two dice. It is easier to compute the probability that this will not happen. This is a case of 24 independent trials. Examining Figure 11.1 we see that the probability of getting a double-six on one throw of two dice is $1/36$, and, hence, the probability of not getting a double-six on one throw is $35/36$. From independence, the probability of not getting double-six on two throws is

$$\left(\frac{35}{36}\right)^2 = \frac{1225}{1296} \approx 0.94522$$

Continuing, independence shows that the probability of never throwing double-six in 24 throws is

$$\left(\frac{35}{36}\right)^{24} \approx 0.50860$$

Consequently, the probability of throwing at least one double-six is approximately

$$1 - 0.50860 = 0.49140$$

Thus we can advise the Chevalier de Méré that game 1 of Problem 11.1 is more favorable to the dice thrower than game 2.

The Birthday Matching Problem

We consider another problem dealing with independence. Many find the answer to the following unexpected.

Problem 11.15. A group of people is selected randomly with respect to their birthdays.[4] If there are 366 people,[5] then it is certain that at least two people share a birthday. How large must the group be to ensure that the probability is greater than one-half that at least two of them share the same birthday?

Solution. We ignore the effect of leap years, and we assume that the probability that a randomly chosen person has a birthday on a particular day is $1/365$; that is, all days are equally likely as birthdays.

This problem deals with n independent trials. We determine an arbitrary order for the trials as follows. If the group contains n people, then we give them consecutive numbers from 1 to n. The sample space consists of n-tuples of the natural numbers 1 through 365.

It is easier to compute the probability that everyone's birthday is different. We compute the probability p_n that the nth person's birthday is different from everyone else's with a number less than n, given that the first $n-1$ people have different birthdays. From independence, the probability that no two people share a birthday is $p_1 p_2 \cdots p_n$.

1. No one has a number less than #1, therefore $p_1 = 1$.

2. The probability that #2's birthday is different from #1's is
$p_2 = 364/365$.

3. The probability that #3's birthday is different from #1's and #2's is
$p_3 = 363/365$.

4. The probability that #4's birthday is different from #1's, #2's, and #3's is
$p_4 = 362/365$.

5. The probability that #n's birthday is different from everyone else's is
$p_n = (365 - n + 1)/365$.

The probability that everyone has a different birthday is the product of these probabilities. For example, the probability that 22 people have different birthdays is

$$\frac{365 \cdot 364 \cdot 363 \cdot 362 \cdots 344}{365^{22}} \approx 0.5243$$

The corresponding probability for 23 people is

$$\frac{365 \cdot 364 \cdot 363 \cdot 362 \cdots 343}{365^{23}} \approx 0.4927$$

It follows that in a random group of 23 people, the probability is slightly greater than one-half that two people share a birthday.

The Method of Inclusion and Exclusion

The algebra of sets leads to a probability problem with a surprising answer—the *affair of the graduation mix-up*—but in preparation we must first deal with a matter of club membership rosters.

In a certain town there is an Angling Club with membership roster A and a Bird-watching Club with membership roster B. We treat A and B as sets. The number of Anglers is $N[A]$ and the number of Bird-watchers is $N[B]$. How many people belong to either club A or club B? The sum $N[A] + N[B]$ is incorrect because this sum counts twice those who belong to both clubs. The correct number is

$$N[A \cup B] = N[A] + N[B] - N[A \cap B] \tag{11.2}$$

I forgot to mention that there is also a Carpentry Club C. How many people belong to at least one of these three clubs? We can reduce this problem to the previous case if we use the fact that the Anglers and Bird-watchers have combined their clubs and now call themselves the Wilderness Club W. Using the previous result we have

$$\begin{aligned} N[A \cup B \cup C] = N[W \cup C] &= N[W] + N[C] - N[W \cap C] \\ &= N[A] + N[B] - N[A \cap B] + N[C] - N[(A \cup B) \cap C] \end{aligned} \tag{11.3}$$

Let's take a closer look at the last term $N[(A \cup B) \cap C]$. This is the number of people who are either Angler-Carpenters or Bird-watcher-Carpenters, that is, the set $(A \cap C) \cup (B \cap C)$. We apply equation (11.2) to obtain

$$N[(A \cap C) \cup (B \cap C)] = N[A \cap C] + N[B \cap C] - N[A \cap B \cap C]$$

Substituting this equality into equation (11.3), we obtain

↩ Dangerous curve!

$$N[A \cup B \cup C] = N[A] + N[B] + N[C]$$
$$- N[A \cap B] - N[A \cap C] - N[B \cap C] + N[A \cap B \cap C] \quad (11.4)$$

Suppose that there are membership lists for each of the clubs, dual membership lists for each pair of clubs, and finally a triple membership list of those who belong to all three clubs. The number of people who belong to at least one of the three clubs is equal to the total of the membership lists of all three clubs *minus* the total of the dual membership lists *plus* the total of the triple membership list. The following fact generalizes this result.

↩ Dangerous curve!

Fact 11.16. *Let A_1, A_2,...,A_n be finite sets. Let N_i $(1 \leq i \leq n)$ be the sum of the sizes of all intersections of exactly i of these sets. Then the number of elements that belong to at least one of the sets is equal to*

> By the *size* of a finite set A we mean the number $N[A]$ of members in the set.

$$N_1 - N_2 + N_3 - N_4 + \cdots \pm N_n \qquad (11.5)$$

Fact 11.16 is to the key to the solution of the following probability problem.

Example 11.17 (the graduation mix-up). Due to an accident, the diplomas were delivered in random order to the graduation ceremony. If there were 100 graduates, what is the probability that at least one graduate got his or her correct diploma?

The solution makes use of concepts and notation defined in Chapter 5.

Solution. The sample space S for this problem consists of all of the 100! possible ways of ordering the 100 diplomas. Let A_i be the subset of S in which the ith student's diploma is in its correct place, ith from the top; A_1 consists of the $(n-1)!$ permutations in which the diplomas other than the ith are ordered arbitrarily. To find N_1 of equation (11.5) we add the sizes of all of the subsets A_i, obtaining

$$N_1 = n(n-1)! = n!$$

Let A_{ij} be the subset of S in which both the ith and jth $(i \neq j)$ diplomas are in the right places. The size of each subset A_{ij} is $(n-2)!$, and there are $\binom{n}{2}$—n choose 2— subsets in all. Thus, we obtain

$$N_2 = (n-2)! \binom{n}{2} = \frac{(n-2)!n!}{2!(n-2)!} = \frac{n!}{2!}$$

In a similar fashion, we have

$$N_3 = \frac{n!}{3!}$$

and, in the general case,

$$N_j = \frac{n!}{j!}$$

Dangerous curve! ↬ The number of permutations in which at least one student gets the right diploma is

$$(N_1 - N_2 + N_3 - N_4 + \cdots - N_{100}) = n! \left(1 - \frac{1}{2!} + \frac{1}{3!} - \frac{1}{4!} + \cdots - \frac{1}{100!} \right)$$

The desired probability of at least one match is equal to

$$1 - \frac{1}{2!} + \frac{1}{3!} - \frac{1}{4!} + \cdots - \frac{1}{100!} \approx 0.63212056 \qquad (11.6)$$

This probability is almost independent of the number of graduates. In fact, this eight-decimal-place approximation remains correct even if there are 1000 graduates. That is, if 1000 graduates are given their diplomas in random order, 0.63212056 is the probability that at least one of them receives his or her correct diploma. It can be shown that the left side of equation (11.6) tends to

$$1 - \frac{1}{e}$$

where e is equal to $2.71828\ldots$.

My school avoids the problem of the graduation mix-up by giving each graduate a blank piece of paper onstage at the ceremony. The actual diplomas are mailed to the graduates at a later date.

Expectation

When the Chevalier de Méré wrote to Pascal, his ultimate concern was how money should be wagered at the gambling table. The following is a simple question of this sort that leads to the important concept of *expectation*.

Problem 11.18. Ada agrees to pay Ben one dollar for each pip that shows on a single throw of one die. To make the game fair, what price should Ada receive from Ben in advance each time they play this game?

Solution. The fair price is the *average* amount per game that Ben receives, but what is meant by average in this instance? We will answer this question in a more abstract setting shortly. A reasonable average is the arithmetic mean of the amounts that would be paid for each of the six equally likely outcomes.

arithmetic mean. The arithmetic mean of n numbers $a_1, a_2, \ldots a_n$ is defined as

$$\frac{a_1 + a_2 + \cdots + a_n}{n}$$

$$\text{Average} = \frac{1 + 2 + 3 + 4 + 5 + 6}{6} = \frac{21}{6} = \$3.50$$

The foregoing problem illustrates the concept of *expectation* that we will define in general, but first we must define what is meant by a *random variable*.

Definition 11.19. A real-valued function f defined on a probability space is called a *random variable*.

In the context of gambling, the amount won or lost is a random variable.

Definition 11.20. Let f be a random variable on a finite probability space S with n points $s_1, s_2, \ldots, s_n$ with probabilities $p_1, p_2, \ldots, p_n$. Then the *expectation* $\mathcal{E}[f]$ of f over S is defined as the sum

$$p_1 f(s_1) + p_2 f(s_2) + \cdots + p_n f(s_n)$$

In all of the finite probability spaces that we have considered so far, the single points are equally likely, which is the same as saying that the p's in Definition 11.20 are all equal. The following fact applies in this special case.

Fact 11.21. *If the n points of the probability space S are equally likely, then*

$$p_1 = p_2 = \cdots = p_n = \frac{1}{n}$$

and the expectation of f over S is the arithmetic mean of the values of f, that is,

$$\mathcal{E}[f] = \frac{f(s_1) + f(s_2) + \cdots + f(s_n)}{n}$$

The following are two important facts about expectation.

Fact 11.22. *Let f and g be random variables on a probability space S. Putting the random variable h equal to the sum $f + g$, we have*

$$\mathcal{E}[h] = \mathcal{E}[f] + \mathcal{E}[g]$$

Fact 11.23. *Let f be a random variable on a probability space S and let c be a real number. Putting the random variable h equal to the product cf, we have*

$$\mathcal{E}[h] = c\mathcal{E}[f]$$

Problem 11.24. On throwing two dice (red and blue), find the expected number of sixes.

Solution. Define the random variable f to be equal to 1 if the *red* die is a six and 0 otherwise. Similarly, define g to be equal to 1 if the blue die is a six and 0 otherwise. The expected number of sixes is the expectation of the random variable $h = f + g$. The expectation of f is $1/6$, which is the same as the expectation of g. Consequently, by Fact 11.22, the expected number of sixes on one throw is $1/6 + 1/6 = 1/3$.

We can confirm the correctness of Fact 11.22 by computing the expectation of h without using Fact 11.22. The sample space of 36 points for this problem is shown in Figure 11.1. There are 10 points in the sample space for which the number of sixes is 1, one point for which the number of sixes is 2, and for all other points the number of sixes is 0. Thus, we have

$$\mathcal{E}[h] = \frac{10}{36} \cdot 1 + \frac{1}{36} \cdot 2 = \frac{1}{3}$$

We conclude that if Ada agrees to give Ben three dollars for each six that appears on one throw of two dice, then to make the game fair Ben should pay her one dollar in advance each time they play.

In the following problem, we study the effectiveness of a gambling system.

Problem 11.25. A coin is tossed three times. At each toss Ben may decide an amount x to bet. If the result of a coin toss is heads, then Ada pays Ben the amount x; if it is tails, then Ben pays Ada the amount x. Find the expected total amount that Ada pays Ben after three tosses if Ben bets one dollar, except that he doubles his bet each time he loses.

First solution. We define three random variables, f, g, and h, the payment following the first, second, and third tosses, respectively. The expectation of each of these payoffs is zero. The total payment is the sum of these three random variables, and by Fact 11.22 the expectation of this sum is zero. This solution is independent of Ben's betting strategy. There is no system that Ben can employ to change his expected payoff from zero.

Second solution. We can also compute directly the expectation of Ben's payoff using the betting strategy described above, that is, doubling the bet after each loss and returning to the one dollar bet after each win. This computation, since it does not use Fact 11.22, is another confirmation of that fact.

Heads/Tails	Payoff	
HHH	$1 + 1 + 1 =$	3
HHT	$1 + 1 - 1 =$	1
HTH	$1 - 1 + 2 =$	2
HTT	$1 - 1 - 2 =$	-2
THH	$-1 + 2 + 1 =$	2
THT	$-1 + 2 - 1 =$	0
TTH	$-1 - 2 + 4 =$	1
TTT	$-1 - 2 - 4 =$	-7

The above table shows Ben's payoff using this strategy for every possible sequence of three tosses of the coin. Since these eight possibilities are equally likely, the required expectation is the arithmetic mean of these eight payoffs, which is equal to

$$\frac{3 + 1 + 2 - 2 + 2 + 0 + 1 - 7}{6} = 0$$

The question of gambling systems has been the subject of some very sophisticated mathematical analysis,[6] but the results obtained offer no comfort to patrons of gambling casinos. The benefit is all to the other side of the table.

The following example, frustration solitaire, is similar to Example 11.17, the graduation mix-up. As often happens, in Example 11.26 computing the expectation is easy, but computing the related probability is difficult.

Example 11.26 (frustration solitaire). Shuffle thoroughly a deck of 52 playing cards. Deal the cards face up. Call out, "Ace," for the first card dealt, "Two," for the second card, and so on. Deal the entire deck in this manner, repeating four times the thirteen ranks, "Ace, Two, ..., Ten, Jack, Queen, King." If a card that you deal matches a rank that you call, then you lose.

1. What is the expected number of matches?

2. What is the probability of winning the game?

Solution of 1. On dealing one card and calling a random rank, the expected number of matches is $1/13$. By Fact 11.22, the expectation of the total number of matches for the entire deck of 52 cards is the sum of the 52 expected matches for each card. Since for each card the expected number of matches is $1/13$, this total is

$$52 \cdot \frac{1}{13} = 4$$

Solution of 2. Like Example 11.17 (the graduation mix-up), this problem can be solved by the method of inclusion and exclusion, but the details[7] of the solution are, surprisingly, far more complex than Example 11.17.

The Petersburg Paradox

The following gambling game was first discussed by Daniel Bernoulli in the *Commentarii* of the Petersburg Academy—hence, the name *Petersburg paradox*.[8]

Problem 11.27. Ada tosses a coin one or more times until heads appears. She agrees to pay Ben one dollar if heads appears on the first toss, two dollars if on the second toss, four dollar if on the third toss, and so on. In general, she agrees that she will pay Ben 2^n dollars if heads appears for the first time on the nth toss. If the game is to be fair, what should Ben pay Ada initially for the privilege of playing the game?

	There were three distinguished Swiss mathematicians with the name Bernoulli:

1. Jacob (1654–1705) (also known as Jacques or James),
2. Jacob's brother Johann (1667–1748) (also known as Jean or John), and
3. Johann's son Daniel (1700–82).

Solution. To make the problem more amenable to the methods that we have developed, let us assume a further stopping rule. For example, let us say that if heads does not appear in one million tosses, then the game is over and Ada owes nothing. The probability that the first heads occurs on the nth toss is $1/2^n$. To compute the expectation we multiply each such probability by the associated payoff of 2^n. Thus, we see that each toss contributes one dollar to the total expectation of the payoff. Since there are one million tosses to consider, the total expected payoff is one million dollars, which is the amount that Ben should pay for the privilege of playing this game.

The paradox is that no rational person would agree to pay one million dollars to play this game. What is wrong?[9]

One of the difficulties is that this game assumes that Ada is in possession of a rather large fortune. In fact, if heads appears for the first time on the one-millionth toss, then she owes Ben the sum of

$$\$2^{10^6} \approx \$10^{301,030}$$

a number that is larger by many orders of magnitude than the number of electrons in the universe.

To bring the problem back into the realm of plausibility we can reduce the basic payoff to one cent and reduce the maximum number of tosses to 25. Then Ada risks losing 2^{25} cents ($\$355,544.32$) in the event that heads occurs for the first time on the twenty-fifth toss, and the fair price for Ben to play the game is reduced to 25 cents.

Ben's decision to pay a small sum for a small chance at a large prize can be compared to buying a lottery ticket.

In this chapter, we have opened the door to probability theory.[10] In this chapter and in the previous one, we have dealt with sets in an abstract setting. In the next three chapters, we will look at sets that are quite familiar in plane geometry. We begin by looking at the first great scientific discovery, the Pythagorean theorem.

Part III

BALANCING

Chapter 12

Cutouts

I'm very well acquainted too with matters mathematical,
I understand equations, both the simple and quadratical,
About binomial theorem I'm teeming with a lot of news—
With many cheerful facts about the square of the hypotenuse.
—W. S. GILBERT, *The Pirates of Penzance*

Geometric dissection is an attractive form of mathematical recreation and a method of proving a very important result, the Pythagorean theorem, which leads to a discussion of Pythagorean triples and Fermat's last theorem.

Geometric Dissections

There is a kind of mathematical puzzle that is concerned with dissecting a geometric figure into a number of pieces and reassembling them to form a different figure. Problems of this sort are called geometric dissections.

For example, in Figure 12.1, an equilateral triangle is cut into four pieces, and it is astonishing that they can be reassembled to form a square. We can demonstrate this fact by making a tracing of Figure 12.1, and cutting out the four pieces, and then reassembling them. This construction is from the British puzzle creator, Henry Dudeney.[1] If the pieces are hinged, then they can be assembled as shown in Figure 12.2. A skilled woodworker could make an interesting set of tables with this design. Figure 12.3 shows two of Dudeney's dissections of a Greek cross into a square.[2]

Figure 12.4 is a further example of a dissection in which four irregular congruent pentagons and a square are assembled to form either an octagon or a square.[3]

Geometric dissections are more than just interesting mathematical recreations. Dissections are used to prove a result that is of fundamental importance in mathematics and hence in all of the sciences that use mathematics. This result—the Pythagorean theorem—can be called the first great scientific discovery.

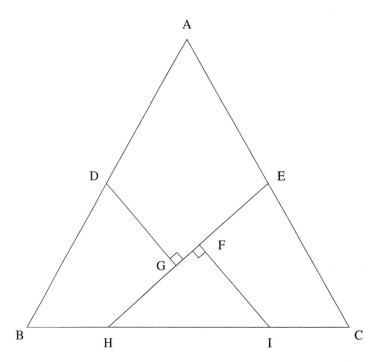

Figure 12.1. This equilateral triangle can be made into a square by rearranging these four pieces as shown in Figure 12.2. The points D and E are the midpoints of the sides AB and AC, respectively. The point H is placed so that the length of EH is equal to the length of the side of a square having the same area as triangle ABC. The point I is placed so that the length of HI is equal to the length of EC—that is, one half the length of a side of triangle ABC. Lines DG and IF are both perpendicular to line EH. The reader may wish to use this diagram as a pattern in order to cut out and reassemble the pieces to form a square.

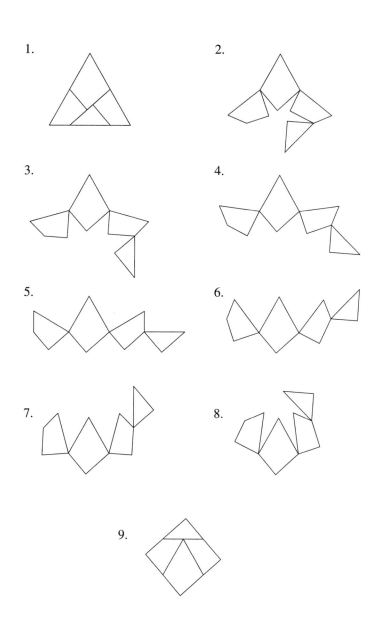

Figure 12.2. Four hinged pieces, also shown in Figure 12.1, can be rearranged to form either an equilateral triangle or a square.

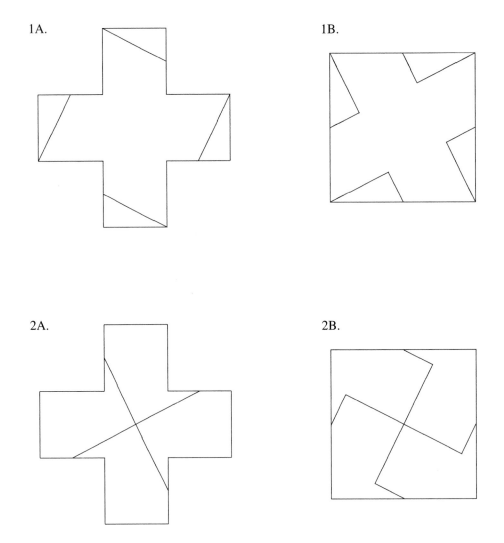

Figure 12.3. Two dissections of a Greek cross into a square.

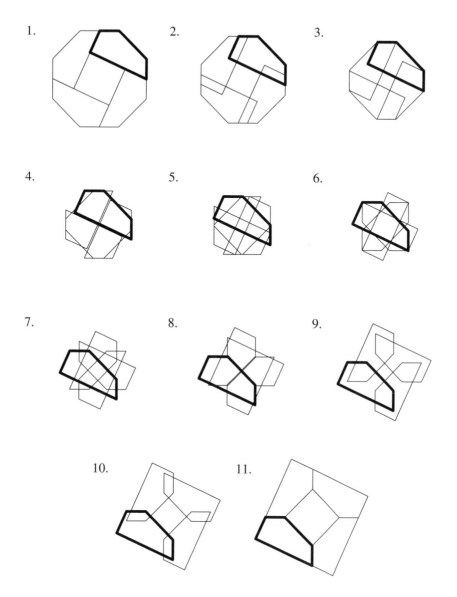

Figure 12.4. Five pieces, four congruent irregular pentagons and a square, can be rearranged to form either an octagon (step 1) or a square (step 11). In the above demonstration, the pentagons move by pairs towards each other and finally, in step 11, change places. The pentagons overlap except in steps 1 and 11. In the transformation process, the small square which is initially between the pentagons in step 1 reappears in step 11 with a small clockwise rotation.

The Pythagorean Theorem

We begin with a statement of the Pythagorean theorem.

hypotenuse. In a right triangle, the side opposite the right angle is called the *hypotenuse*. The other two sides are called *legs*.

Theorem 12.1 (the Pythagorean theorem). *In a right triangle, let a and b be the length of the two legs and let c be the length of the hypotenuse. Then $a^2 + b^2 = c^2$.*

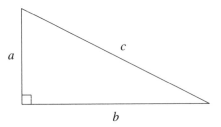

Figure 12.5. The Pythagorean theorem: $a^2 + b^2 = c^2$.

Euclid's theorem on the infinitude of the prime numbers[4] may equal or exceed the Pythagorean theorem in elegance, but the latter theorem is more important. The reason is that the Pythagorean theorem is—to a greater extent than Euclid's theorem—a stepping-stone to other mathematical results. The Pythagorean theorem is useful to surveyors and draftsmen, but even more important are the generalizations of the Pythagorean theorem in mathematical *abstract spaces*—structures with many scientific applications outside mathematics.

There are a great many proofs of the Pythagorean theorem, and many of them involve geometric dissections. In fact, E. S. Loomis[5] collected 370 proofs of this theorem. One of the proofs is from the American president, John Abram Garfield (1831–81). We might add that Garfield made his discovery 2500 years after Pythagoras died. Since we will see four very different proofs, it will become plausible that there are many other proofs. First, we will examine the dissection proof that some believe was discovered by Pythagoras; then we will look at President Garfield's proof; and, finally, we will see two proofs from the twelfth-century Indian mathematician Bhāskara.

The proof attributed to Pythagoras

No one knows for sure what method Pythagoras used. The following proof is based on Figure 12.6, which consists of two $a + b$ by $a + b$ squares subdivided differently. In the left diagram, the large square is dissected into four right triangles and two squares, one square of area a^2 and the other of area b^2. In the right diagram, the large square is subdivided into four right triangles—the same four as before but in different positions—and one square of area c^2. Furthermore, it is clear that the total area of both diagrams is the same. Subtracting the areas of the four right triangles

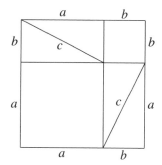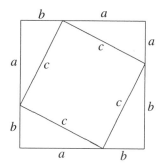

Figure 12.6. Dissections for proof of the Pythagorean theorem.

from the total area of the large squares we see that the combined area of the two squares on the left $(a^2 + b^2)$ is equal to the area c^2 of the c by c square on the right. We know that this figure is a square because of its obvious symmetry. In fact, since the figure is unchanged after a rotation of $90°$, it follows that the four angles of the quadrilateral are equal; therefore, it must be a square.

There are many other proofs of the Pythagorean theorem that are dissections, but now let us look at two proofs of a different sort.

President Garfield's proof

President Garfield's proof is not a dissection; it does not involve cutting out figures and rearranging them. His proof makes use of the formula for the area of a trapezoid. A trapezoid is a quadrilateral with at least one pair of parallel sides. The area of the trapezoid shown in Figure 12.7 is $h(a+b)/2$, but we shouldn't digress to prove this formula. Suffice it to say that this formula for the area can be proved without first assuming that the Pythagorean theorem is true.

Figure 12.8 is the basis of President Garfield's proof. The crux of the matter is that three right triangles can be assembled to form this trapezoid. Two of the triangles are the same right triangles that appeared in the previous dissection (Figure 12.6). The third triangle is an isosceles right triangle. It is not immediately clear that the isosceles triangle is necessarily a *right* triangle. This can be proved by using the fact that the sum of the angles of a triangle is always $180°$.

isosceles. A triangle is said to be isosceles if it has two sides of equal length.

President Garfield's proof proceeds by observing that there are two different ways to compute the area of the trapezoid in Figure 12.8.

1. The area can be computed using the formula for the area of a trapezoid. The distance between the parallel sides is $a + b$, hence the area is

$$\frac{1}{2}(a+b)(a+b) = \frac{1}{2}(a^2 + 2ab + b^2)$$

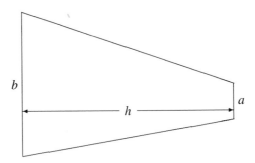

Figure 12.7. A trapezoid is a quadrilateral with two parallel sides. In this figure, the parallel sides have lengths a and b, and the distance between the parallel sides is h. The formula for the area of a trapezoid is $h(a+b)/2$.

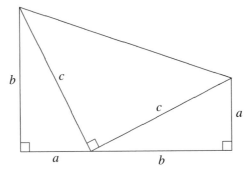

Figure 12.8. President Garfield's proof of the Pythagorean theorem is based on a trapezoid with parallel sides of length a and b, respectively, and with base of length $a+b$.

    2. The area can also be computed by adding the areas of the three triangles, which gives

$$2 \cdot \frac{1}{2}ab + \frac{1}{2}c^2$$

By equating these two expressions that both represent the same area, we obtain

$$\frac{1}{2}(a^2 + 2ab + b^2) = 2 \cdot \frac{1}{2}ab + \frac{1}{2}c^2$$

which implies

$$a^2 + b^2 = c^2$$

Bhāskara's proofs

In this section, we consider two proofs by the Indian mathematician and astronomer Bhāskara (1114–c. 1185).

Bhāskara's dissection proof

The first proof is a dissection proof. Instead of giving a proof in the normal sense, Bhāskara simply exhibits Figure 12.9 with the single word, "Behold!" The proof is not quite that simple. The area of the large square on the left is c^2. The dissection shows that the area of the large square on the left must be equal to the sum of the areas on the right of two $a \times b$ rectangles and a square with area $(b-a)^2$. The Pythagorean theorem follows from the calculation

$$c^2 = 2ab + (b-a)^2 = a^2 + b^2$$

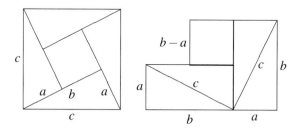

Figure 12.9. Behold! Bhāskara's dissection proof.

Since Bhāskara's second proof uses *similar triangles*, we will digress briefly to consider this topic, which is of independent interest.

Similar triangles

Definition 12.2. Triangles ABC and A′B′C′ are said to be *congruent* if corresponding sides are of equal length.[6]

Definition 12.3. Triangles ABC and A′B′C′ are said to be *similar* if corresponding sides are proportional.

Referring to Figure 12.10, that corresponding sides are proportional means that there exists a positive constant k such that

$$a = ka' \quad b = kb' \quad c = kc'$$

This fact may also be expressed without mentioning the factor of proportionality k as follows.

$$\frac{a}{a'} = \frac{b}{b'} = \frac{c}{c'}$$

The following equalities are consequences of the above equalities.

$$\frac{a}{b} = \frac{a'}{b'} \qquad \frac{a}{c} = \frac{a'}{c'} \qquad \frac{b}{c} = \frac{b'}{c'} \tag{12.1}$$

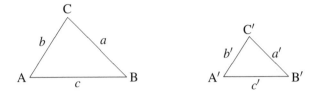

Figure 12.10. Similar triangles.

To determine that a given pair of triangles are similar, the following facts are useful.

Fact 12.4. *Two triangles are similar if and only if corresponding angles are equal.*

Fact 12.5. *Two triangles are similar if corresponding sides are parallel.*

Fact 12.6. *Two triangles are similar if corresponding sides are perpendicular.*

Bhāskara's proof using similar triangles

Bhāskara's proof proceeds as follows. In Figure 12.11, the perpendicular of length d divides the right triangle with sides a, b, c into two smaller triangles that, by Fact 12.4, are similar to the large triangle. From equation (12.1) we have

$$\frac{c}{b} = \frac{b}{m} \qquad \frac{c}{a} = \frac{a}{n}$$

These equations are equivalent to

$$cm = b^2 \qquad cn = a^2$$

Adding these two equations, we obtain

$$a^2 + b^2 = c(m+n) = c^2$$

which is the Pythagorean equality.

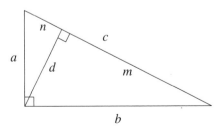

Figure 12.11. Bhāskara's proof of the Pythagorean theorem.

Double-angle triangles

The following is a further application of similar triangles. A *double-angle triangle* is a triangle in which one angle is twice one of the other angles. Figure 12.12 gives an example of a double-angle triangle. As shown in the caption, triangle ANB is similar to triangle ABC by Fact 12.4. From equations (12.1), we have

$$\frac{c}{b} = \frac{b-d}{c} \qquad \frac{b}{2a} = \frac{c}{d}$$

From these two equations we obtain

$$c^2 = b(b-d) = b^2 - bd \tag{12.2}$$

and

$$bd = 2ac \tag{12.3}$$

Use equation (12.3) to eliminate bd from equation (12.2), obtaining

$$c^2 = b^2 - 2ac \tag{12.4}$$

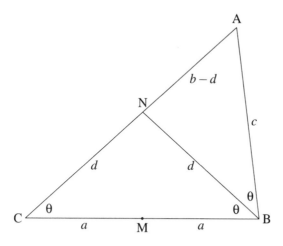

Figure 12.12. A double-angle triangle. Angle ABC is twice the angle ACB, which is equal to θ. Side CB has length $2a$ with midpoint M. Side AC has length b. Point N is the intersection of the bisector of angle ABC with side AC. Therefore, triangle CBN is isosceles with sides CN and BN both of length d. Triangle ANB is similar to triangle ABC by Fact 12.4 because corresponding angles are equal; in fact, the two triangles share angle CAB and angles ACB and ABN are both equal to θ, and since two corresponding pairs of angles are equal, so is the third pair because the sum of the angles of a triangle must be 180°.

Equation (12.4) is a relationship between the lengths of the three sides that is satisfied by the double-angle triangles. For example, for the particular triangle shown in
Figure 12.12, the lengths a, b, and c happen to be proportional to the numbers 5, 12,
and 8. It is easy to verify equation (12.4) in this particular case:

$$c^2 = 8^2 = 64$$
$$b^2 - 2ac = 12^2 - 2 \cdot 5 \cdot 8 = 144 - 80 = 64$$

Equation (12.4) for double-angle triangles is like the Pythagorean theorem (Theorem 12.1) for right triangles. This similarity goes deeper than mere appearance for
the following reason. By adding $2ac + a^2$ to both sides of equation (12.4) we obtain

$$c^2 + 2ac + a^2 = a^2 + b^2$$

But notice that the left side is a perfect square, so that the equation can be written

$$(c + a)^2 = a^2 + b^2 \tag{12.5}$$

Now the connection to the Pythagorean theorem is complete because equation (12.5)
means that there exists a right triangle with hypotenuse $c + a$ and legs a and b. In
fact, imagine that in Figure 12.12, the vertexes A, B, and C, and the midpoint M
are equipped with hinges. Figure 12.13 shows how this hinged quadrilateral can be
deformed into a right triangle. Note that throughout this transformation, the lengths
a, b, and c remain unchanged.   One can ask the question, Does equation (12.4)
characterize the double-angle triangles? In other words, given numbers a, b, and c
satisfying equation (12.4), is there a double-angle triangle with the given numbers
equal to the lengths of the sides as shown in Figure 12.12? No, because in an actual
double-angle triangle the sum of the lengths of sides AB and AC must exceed the
length CB; in other words, we must also have the inequality

$$b + c > 2a \tag{12.6}$$

We can see that there exist numbers a, b, and c that satisfy equation (12.4) but do not
satisfy inequality (12.6)—for example

$$a = 20 \quad b = 21 \quad c = 9$$

Inequality (12.6) is equivalent to $b < 2c$ in the following sense.

Fact 12.7. *Let positive numbers a, b, and c satisfy equation* (12.4). *Then inequality*
(12.6) *is satisfied if and only if b and c satisfy the following inequality:*

$$c < b < 2c \tag{12.7}$$

The following fact asserts that equation (12.4) together with inequality (12.7)
characterize the double-angle triangles.

Fact 12.8. *Positive numbers a, b, and c are the lengths of the sides of a double-angle*
triangle if and only if both equation (12.4) *and inequality* (12.7) *are satisfied.*

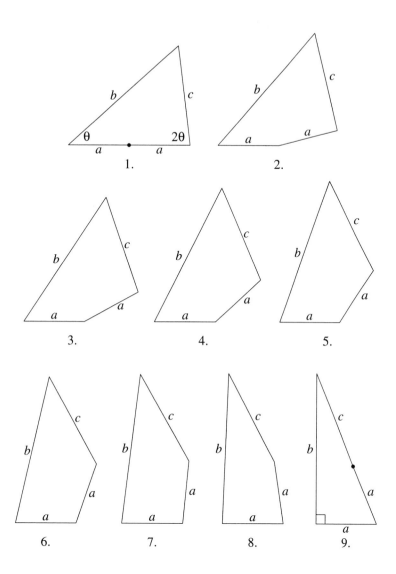

Figure 12.13. A hinged quadrilateral can be deformed into either a double-angle triangle or a right triangle.

Every double-angle triangle can be deformed into a right triangle by the process shown in Figure 12.13. However, because of Fact 12.8, the right triangle in Figure 12.13.9 can be deformed into the double-angle triangle in Figure 12.13.1 only if we have the additional condition $b < c < 2b$.

For *right triangles*, the situation is simpler than it is for double-angle triangles because right triangles are characterized by the Pythagorean equality of Theorem 12.1 without an additional condition like inequality (12.7). In the following section, we discuss right triangles in which the lengths of all three sides are integers.

Pythagorean Triples

From the formula

$$3^2 + 4^2 = 9 + 16 = 25 = 5^2$$

> That the sides of a triangle are *proportional* to 3, 4, and 5 means that there is a positive number k (not necessarily an integer) such that the actual lengths of the sides are $3k$, $4k$, and $5k$.

it follows from the Pythagorean theorem that a triangle with sides proportional to 3, 4, and 5 must be a right triangle—a fact that was known by the ancient Babylonians. It is possible to construct an accurate right angle by forming such a 3, 4, 5 triangle using carefully measured pieces of rope.

It is an interesting problem to find other *Pythagorean triples*, that is, to find other triples of natural numbers a, b, and c such that $a^2 + b^2 = c^2$. There is no need to carry out the search by trial and error because there is a more efficient method that we will state in the form of a proposition.

Proposition 12.9. *Let m and n be arbitrary natural numbers. Then the integers*

$$x = m^2 - n^2 \quad y = 2mn \quad z = m^2 + n^2$$

satisfy the equation

$$x^2 + y^2 = z^2 \tag{12.8}$$

Proof. We calculate

$$x^2 = \left(m^2 - n^2\right)^2 = m^4 - 2m^2n^2 + n^4$$
$$y^2 = (2mn)^2 \quad = \quad \quad 4m^2n^2$$
$$z^2 = \left(m^2 + n^2\right)^2 = m^4 + 2m^2n^2 + n^4$$

from which equation (12.8) follows. $\qquad\qquad\qquad\qquad\qquad\qquad \square$

Thus, we see that there are infinitely many Pythagorean triples. Figure 12.14 shows a few Pythagorean triples.

This is essentially the only way to obtain Pythagorean triples. However, for each Pythagorean triple x, y, z obtained from Proposition 12.9, kx, ky, yz is also a Pythagorean triple for every natural number k, and *all* Pythagorean triples are generated in this way.[7]

m	n	x	y	z
2	1	3	4	5
3	2	5	12	13
4	1	15	8	17
4	3	7	24	25
5	2	21	20	29
5	3	16	30	34
5	4	9	40	41
6	1	35	12	37
6	5	11	60	61

Figure 12.14. Pythagorean triples.

Double-angle triangles with integer sides

Proposition 12.9 together with Fact 12.8 gives a way of finding double-angle triangles with integer sides. Specifically, referring to Figure 12.12, put

$$a = x = m^2 - n^2 \quad b = y = 2mn \quad c = z - x = m^2 + n^2 - (m^2 - n^2) = 2n^2$$

According to Fact 12.8, these values for a, b, and c give double-angle triangles with integer sides provided that inequality (12.7) also holds. This translates to the following inequality for the integers m and n.

$$2n^2 < 2mn < 4n^2$$

This inequality simplifies to

$$n < m < 2n$$

Figure 12.15 lists a few double-angle triangles with integer sides. The triangle $a = 5, b = 12, c = 8$ is shown in Figures 12.12 and 12.13.1.[8]

m	n	a	b	c
3	2	5	12	8
4	3	7	24	18
5	3	16	30	18
5	4	9	40	32
6	5	11	60	50

Figure 12.15. Double-angle triangles with integer sides.

The Pythagorean triples lead to a very famous problem that has recently been solved.

Fermat's last theorem

The French mathematician Pierre de Fermat owned a copy of *Arithmetica*, a work of Diophantus of Alexandria. The exact dates of Diophantus are not known, but he is thought to have lived in the third century A.D. Diophantus studied the problem of finding integer solutions to equations. In fact, problems of this sort are called *diophantine equations*. For example, Proposition 12.9 is concerned with the diophantine equation (12.8). In Book II, Problem 8, of *Arithmetica*, Diophantus asks for integer solutions of equation (12.8) as follows. He asks the reader, "to decompose a given square number into the sum of two squares," to which Fermat added in the margin, "However, it is impossible to write a cube as the sum of two cubes, a fourth power as the sum of two fourth powers and in general any power beyond the second as the sum of two similar powers. For this I have discovered a truly wonderful proof, *but the margin is too small to contain it.*" More formally, we state this as follows.

Theorem 12.10 (Fermat's last theorem). *There do not exist natural numbers x, y, z, and n such that n is greater than 2 and*

$$x^n + y^n = z^n \tag{12.9}$$

It is interesting to note that it is easy to show that for a slight modification of equation (12.9) there are infinitely many solutions for each value of n. The modified equation is

$$x^n + y^n = z^{n+1} \tag{12.10}$$

Proposition 12.11. *For each natural number n, there exist infinitely many integer solutions of equation* (12.10).

Proof. Let a and b be natural numbers. It is an easy algebraic exercise to see that we have

$$[a(a^n + b^n)]^n + [b(a^n + b^n)]^n = (a^n + b^n)^{n+1}$$

Therefore, we obtain a solution to equation (12.10) by putting

$$x = a(a^n + b^n)$$
$$y = b(a^n + b^n)$$
$$z = a^n + b^n \qquad \qquad \square$$

Fermat's marginal note was discovered only after his death, and we will never know for sure whether he did discover a proof. We do know that for more than 300 years many prominent mathematicians have unsuccessfully attempted to find a proof. As a by-product of their research, they discovered a great deal of new mathematics. The problem received an added impetus in 1908, when a mathematician at the University of Göttingen bequeathed the sum of 100,000 German marks to be given for the first proof of Fermat's last theorem. That prize became greatly devalued in the German hyperinflation of 1923, but that did not diminish the interest in this problem.

There was much that argued for the truth of Fermat's last theorem. It was known that Fermat's last theorem is true provided that n is smaller than a certain very large number. But the clincher came in 1994 when mathematician Andrew Wiles at Princeton announced a proof for Fermat's last theorem. It turned out that there was a flaw in his proof, but, in 1995, Wiles published a revised proof that is now accepted as correct by experts in the field. After more than 300 years of intense mathematical effort, Fermat's last theorem has finally been proved.

This chapter is concerned with geometric dissections and the Pythagorean theorem. In the next chapter, we will see geometric applications of the lever.

Chapter 13

Seesaw, Marjorie Daw

Eudoxus and Archytas had been the first originators of this far-famed and highly prized art of mechanics, which they employed as an elegant illustration of geometrical truths.

—PLUTARCH (A.D. c. 46–c. 120), *Lives*

The lever is a practical tool of the workaday world, but, as we will see in this chapter and the next, it is also an intellectual tool in the world of mathematics.

The Lever

A seesaw is a long plank balanced on a central support called a *fulcrum* so that when there is a child on each end, one end goes up as the other goes down. It is possible for the seesaw to be in equilibrium even though two different-sized children are riding it. For the seesaw to be balanced, there must be a certain relationship between the weights of the two children and their distances from the central fulcrum. In mechanics, the seesaw is called a *lever*.

It is useful to approximate the problem by making it more abstract. Let's ignore the distribution of mass in the plank, and assume that, instead of children, two unequal weights are placed on the lever.

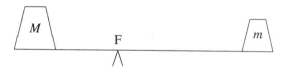

In fact, let's make it even more abstract. Let's suppose that the entire mass is concentrated into just two points on a horizontal line. As we will see, many calculations are unaffected by replacing the mass of a body by the total mass concentrated at a single point called the *center of mass*—sometimes also called the center of gravity. (Figure 13.1.)

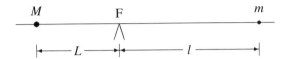

Figure 13.1. The lever with the masses M and m is balanced if $LM = lm$ where L and l are the respective horizontal distances from the fulcrum F to the centers of mass. This is equivalent to the assertion that F is the center of mass of the system consisting of the masses M and m.

> **mechanics.** Mechanics is the branch of physics concerned with forces and motion (i.e., **dynamics** and **kinematics**).
>
> **point mass.** In mechanics, a mass concentrated at a single point is called a *point mass*.

When point masses are placed on a straight line, and a certain point F is designated as the fulcrum, as is the case with the lever, then each mass contributes a quantity that we call its *moment* with respect to the fulcrum. A mass point m located a distance l from the fulcrum contributes a moment equal to ml, the product of the mass times the distance. The moments of all the mass points are added to obtain the total moment.

Mechanics tells us that a system of masses is in equilibrium with respect to rotation about the fulcrum if the total moment is zero, provided that for the distances we take the *signed* distances to the fulcrum—distances that can be either positive or negative. We define signed distances by choosing a positive direction on the line. A line with a positive direction defined is called a *directed line*. Usually the positive direction is taken to be to the right. If this is done for the lever, then the distance of the child on the left from the fulcrum is negative. However, in Figure 13.1 the letter L signifies the distance in the ordinary sense, the positive distance. Therefore, the moment contributed by the child on the left is $-ML$, and the total moment is $ml - ML$. It follows that the condition for equilibrium is $ml = ML$.

Definition 13.1. If n masses, $m_1, \ldots, m_n$, are placed on a directed line with respective *signed* distances from the fulcrum F equal to $l_1, \ldots, l_n$, then the *total moment* with respect to F is

$$m_1 l_1 + \cdots + m_n l_n$$

Suppose that the positive direction is by definition to the right, and suppose that the total moment is positive. Then Newton's laws of motion tell us that if the total moment is positive, then, under a downward force of gravity, the lever, initially at rest, will rotate clockwise. This is reasonable because clearly a large weight placed on the right side of the lever will produce a positive moment and will cause clockwise rotation. Similarly, if the total moment is negative, then the lever, initially at rest, will rotate counterclockwise.

The following fact and definition deal with the one-dimensional mass distributions that we are discussing, but they can be generalized to two and three dimensions.

Fact 13.2. *The necessary and sufficient condition for equilibrium with respect to a point F is that the total moment with respect to F is equal to zero.*

Definition 13.3. If a system of mass points is in equilibrium with respect to a fulcrum point F, then F is also called the *center of mass* of the system.

Given a distribution of point masses on a line, we can find the center of mass as follows. Let the masses be $m_1, \ldots, m_n$ and let $x_1, \ldots, x_n$ be the respective signed distances from a particular point O called the origin. If $\bar{x}$ is the center of mass, then $x_i - \bar{x}$ is the signed distance from the ith mass to the center of mass. The total moment with respect to the center of mass must be zero. In other words,

$$(x_1 - \bar{x})m_1 + \cdots + (x_n - \bar{x})m_n = 0$$

For $\bar{x}$ we say, "x bar."

It is not hard to solve this equation for $\bar{x}$.

$$\bar{x} = \frac{x_1 m_1 + \cdots + x_n m_n}{m_1 + \cdots + m_n} \tag{13.1}$$

Notice that the numerator is the total moment with respect to the origin and the denominator is the total mass of the system.

The most important fact about the center of mass is that for many purposes a system of mass points behaves as though the total mass is concentrated at the center of mass. For this discussion, let us denote a mass point on a directed line as (m, x) where m is the mass and x is the directed distance of the point mass from the origin.

For example, suppose there are two different mass distributions on a line containing p and q points, respectively:

$$(m_1, x_1), (m_2, x_2), \ldots, (m_p, x_p) \tag{13.2}$$

and

$$(M_1, X_1), (M_2, X_2), \ldots, (M_q, X_q) \tag{13.3}$$

Suppose that the distribution (13.2) has center of mass $\bar{x}$ and that (13.3) has center of mass $\overline{X}$. It turns out that the combined system containing all the mass points of both (13.2) and (13.3) resembles a system containing just two points: a mass m equal to the total mass of (13.2) located at $\bar{x}$ and a mass M equal to the total mass of (13.3) located at $\overline{X}$. From (13.1) and some algebra, one can prove the following two surprising facts.

1. The combined distribution, (13.2) together with (13.3), and the two-point distribution $(m, \bar{x}), (M, \overline{X})$ have the same center of mass.

2. The combined distribution and the two-point distribution have the same moment with respect to an arbitrarily chosen fulcrum point.

Both principles can be generalized to a system of point masses that is composed of any finite number of subsystems.

A center of gravity with similar properties is defined for mass distributions in two and three dimensions. Each of the children on the seesaw can be viewed as a very large number of mass points placed in three-dimensional space. To decide if the seesaw is balanced, we only need to know the total mass and center of gravity of each child.

The Leaning Tower of Bricks

We can see instances of equilibrium elsewhere in the playground. For example, suppose that a child in a sandbox is making a tower of brick-shaped blocks. How far can a balanced tower of bricks overhang its base?

The surprising answer is that it is theoretically possible to stack a tower of bricks to overhang its base *as far as we please*. This claim does not conflict with the fact that a tower of real bricks cannot overhang its base more than a few inches because we are speaking of mathematically "perfect" bricks stacked on a mathematically "perfect" table. We will construct a plausible mathematical model for the bricks that permits unlimited overhangs, but the mathematical model is only an approximation for the real world. The extent to which actual bricks conform to the ideal is limited by the precision of the bricks. Another practical consideration is that the number of bricks required for an overhang of 100 feet exceeds the number of electrons in the universe! Despite practical considerations, no matter how far a tower of bricks overhangs its base, there is no theoretical barrier that prevents us from extending the tower farther still.

Before we get into details, we must understand what we mean by a "tower of bricks." Let us agree that a tower of bricks consists of a number of bricks placed one on top of the other so that at any horizontal level there is no more than one brick. Each brick is box-shaped. To make matters specific, let us say that each brick has the dimensions

$$7\frac{1}{2}'' \times 3\frac{1}{2}'' \times 2\frac{1}{4}''$$

We designate arbitrarily on each brick one of the two largest faces $(7\frac{1}{2}'' \times 3\frac{1}{2}'')$ as the *base* and the larger of the two midsized faces $(7\frac{1}{2}'' \times 2\frac{1}{4}'')$ as the *front* of the brick. We assume that in a tower of bricks, the bases of the bricks are all horizontal and that the fronts are all in one vertical plane—as they are in a normal brick wall. In other words, a tower of bricks might look like this.

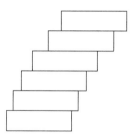

The centers of gravity of the bricks are located at the geometric centers of the bricks. Let us mark the centers of gravity with a dot.

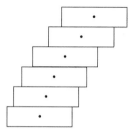

Furthermore, to describe these towers quantitatively, in Figure 13.2 we label the distance that each brick overhangs the brick beneath it.

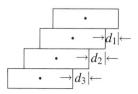

Figure 13.2. A tower of bricks can be specified quantitatively by listing the successive offsets d_1, d_2, d_3.

It is convenient to use half a brick length $(3\frac{3}{4}'')$ as the unit of length. The reason for this choice will appear later.

There is a certain method of stacking bricks optimally to obtain the greatest possible overhang. The *maximal tower* with n bricks, shown in Figure 13.3, will be denoted T_n. The tower T_n is characterized by the offsets

$$d_1 = 1 \quad d_2 = \frac{1}{2} \quad d_3 = \frac{1}{3} \ldots d_{n-1} = \frac{1}{n-1} \tag{13.4}$$

There are two remarkable facts about the maximal tower T_n.

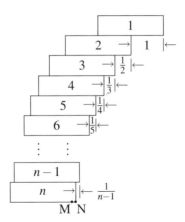

Figure 13.3. The maximal tower T_n of n bricks stacked for maximum overhang. For each integer r $(0 < r < n)$ the rth brick overhangs the $r+1$st brick by $1/r$ where the unit of distance is half a brick's width. Point M is the vertical projection of the center of mass of T_n onto the base of the bottom brick. Point N is the bottom right corner of the bottom brick.

1. The tower T_n is balanced. (See Proposition 13.6.) However, it is just barely balanced—it is in *unstable* equilibrium—because the slightest movement will cause the tower to fall. But, theoretically, if the tower is initially at rest, then the downward force of gravity will not cause it to fall. As a practical matter, if we wish to build a balanced tower, we must use offsets slightly smaller than those shown in Figure 13.3.

2. If we take n large enough, the tower of Figure 13.3 overhangs its base as far as we please. (See Proposition 13.7.)

In the following two sections we deal with these matters separately.

The maximal tower is balanced

We wish to show that the tower T_n is balanced, but before we can proceed, we need to know how to determine whether a stack of bricks is balanced. There is a connection between the lever problem and the brick tower problem. Let's begin by looking at a tower of just two bricks. (Figure 13.4.) Think of the upper brick as a lever with a fulcrum at some point on the surface of the brick below it.

In Figure 13.4(a), the center of mass of the upper brick projects vertically to a point F on the surface of the upper brick, and we consider this to be a fulcrum point. (Of course, what we call the fulcrum point in our diagram is actually a line segment that projects to a point in Figure 13.4.)

It appears to be a difficulty that in Figure 13.4(a) the upper brick is not free to rotate about the fulcrum F. However, consider what happens if, in our imagination, we chisel away all of the lower brick in Figure 13.4(a) except for a vertical knife edge containing F—a process of elimination. Then the upper brick becomes a lever

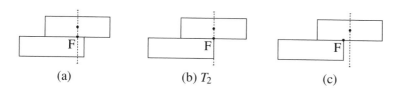

(a)   (b) T_2 (c)

Figure 13.4. In (a) and (b), the upper brick is balanced. In (c), it is unbalanced. Note that (b) is T_2—Figure 13.3 with $n = 2$. The equilibrium in (b) is said to be unstable because even if the upper brick is initially at rest, a suitably placed downward force, however small, will cause it to fall. However, if no force is applied, the upper brick will remain at rest.

with a fulcrum at F. Moreover, the total moment is zero because F is defined as the vertical projection of the center of mass of the upper brick. This means that even the remaining knife edge of the lower brick, assuming that it is strong enough and does not crumble, supports the upper brick in equilibrium—and all the more so if we restore the entire lower brick.

The same analysis applies to Figure 13.4(b) where the upper brick is also balanced. Note that Figure 13.4(b) is T_2—the special case of Figure 13.3 with $n = 2$. However, in Figure 13.4(c), the vertical dotted line through the center of mass of the upper brick misses the lower brick altogether. For this reason, the foregoing argument, the "process of elimination," fails. In fact, we can show that in this case the upper brick topples by choosing the fulcrum F to be the right edge of the lower brick. With this choice for F, the total moment with respect to F in Figure 13.4(c) is positive because the positive direction is to the right, and the horizontal distance between F and the center of mass is not zero.

As we stated before, when the total moment is positive, the lever tends to rotate clockwise. Because there is no impediment to clockwise rotation, the upper brick falls off the lower brick.

Definition 13.4. By a *subtower* of a tower of bricks, we mean the part of the tower that remains when all the bricks below a certain level are removed.

This argument holds equally well if we replace the upper brick with a subtower of bricks—the subtower excludes the bottom brick (Figure 13.5). We are concerned only with the state of equilibrium of the juncture between the subtower and the bottom brick; for the moment, we can assume that all the other bricks are glued together. We have equilibrium if and only if the center of mass of the subtower projects vertically onto the upper surface of the bottom brick.

The normal way of building a tower of bricks is from the bottom up, but the special role of the bottom brick in Figure 13.5 suggests that theoretically it might be useful to build a tower from the top down. The reason for proceeding in this strange fashion is as follows. Suppose that a tower of bricks is in equilibrium. If we add another brick at the top, the equilibrium could fail at the juncture between the top brick and the one below it, but it also might fail at any of the other junctures. However, if a tower is initially balanced, then if we add a brick at the bottom, the

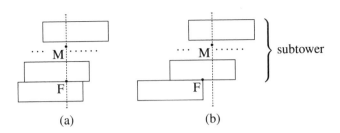

Figure 13.5. A subtower consisting of an indefinite number of bricks rests on the bottom brick. Bricks not shown are indicated by the horizontal series of dots. Point M denotes the center of mass of the subtower. In (a), as in Figure 13.4(a), the fulcrum point F is defined as the vertical projection of M onto the upper surface of the bottom brick, and in (b), as in Figure 13.4(c), the fulcrum point F is defined as the right edge of the bottom brick. In (a) the subtower is balanced on the bottom brick, and in (b) the subtower topples.

only place for equilibrium to fail is at the juncture between the bottom brick and the one above it. We have the advantage that when we add a new brick, we don't have to adjust any of the old bricks.

Thinking about the tower from top down gives us useful insights, even if no one recommends *building* a tower in that way. We can summarize what we have learned so far as follows.

Proposition 13.5. *A tower of bricks is balanced if and only if both of the following conditions hold.*

1. *The subtower formed by removing the bottom brick is balanced.*

2. *The center of mass of this subtower projects vertically onto the horizontal surface of the bottom brick.*

From Proposition 13.5 it follows that either a tower is unbalanced or it is balanced together with all of the subtowers formed by successively removing one brick at a time from the bottom of the initial tower—a process that ends with a balanced tower consisting of a single brick.

We have seen that the tower T_2 is balanced. We are now ready to prove the main result of this section.

Proposition 13.6. *For each natural number n, the tower T_n is balanced. Moreover, the center of mass of T_n projects vertically to a point at distance $1/n$ from the lower right corner of the bottom brick of T_n. In other words, referring to Figure 13.3, the distance $\overline{MN}$ is equal to $1/n$.*

Proof. We proceed by *mathematical induction*. Recall that we used mathematical induction in our study of triangular numbers on page 58. The proof proceeds by two steps.

1. Show that the proposition is true for $n = 1$. Note that T_1 consists of just one brick. The proposition in this case merely asserts the obvious fact that a tower consisting of one brick is balanced and that its center of mass is at the center of the brick. (The unit of distance is half a brick width.)

2. We suppose that for arbitrarily fixed n the proposition is true for the tower T_{n-1}. This assumption—the inductive hypothesis—consists of two parts. (1) We assume that T_{n-1} is balanced. (2) We assume that the center of mass of T_{n-1} projects vertically to a point at distance $1/(n-1)$ from the lower right corner of the $n-1$st brick. From these assumptions we must show that the proposition is also true for T_n—an assertion that, similarly, consists of two parts. (a) We must show that T_n is balanced. (b) We must show that the center of mass of T_n projects vertically to a point at distance $1/n$ from the lower right corner of the nth brick.

   (a) We show that T_n is balanced by showing that conditions (1) and (2) of Proposition 13.6 are satisfied.

      To show condition (1) we must be able to say that the subtower T_{n-1} is balanced; but this is simply a part of the inductive hypothesis.

      We show condition (2) as follows. By the inductive hypothesis, we may assume that T_{n-1} is balanced; therefore, the center of mass of T_{n-1} projects vertically to a point on the base of the $n-1$st brick which is at a distance $1/(n-1)$ from its lower right corner. Referring to Figure 13.3 and noting that the offset between the nth and the $n-1$st brick is defined to be $1/(n-1)$, we see that the center of mass of T_{n-1} projects vertically to the upper right corner of the nth brick, and we take this point to be our origin for the purpose of computing the center of mass of T_n. We make this computation according to equation (13.1) by dividing the total moment by the total mass. Furthermore, for this computation we may consider the mass of T_{n-1} to be concentrated at its center of mass. Since the center of mass of T_{n-1} coincides with the origin, the contribution of $\hookleftarrow$ Dangerous curve! T_{n-1} to the total moment is zero. Therefore, T_n is balanced.

   (b) Now we show that the center of mass of T_n projects vertically to a point at distance $1/n$ from the lower right corner of the nth brick. We take the unit of mass equal to the mass of one brick and the unit of length to be half a brick width. The mass of the nth brick can be assumed to be concentrated at its center of mass, which is one unit to the left of our origin. Thus, the total moment is equal to $1 \times 1 = 1$. The total mass of T_n is n. Hence the center of mass is located, as claimed, $1/n$ to the left of the origin.

This completes the proof by mathematical induction of Proposition 13.6. $\square$

We have shown that for every n the maximal tower T_n is balanced. In the next section we will see that upon choosing n suitably large the maximal tower overhangs as far as we please.

The harmonic series diverges

These offsets (13.4) of T_n are terms in an infinite series

$$1 + \frac{1}{2} + \frac{1}{3} + \cdots$$

which is called the *harmonic series*. (Recall that on page 14 formula (2.1) is another example of an infinite series.) Related to the harmonic series, we define, for each natural number n, the finite sums

$$H_n = 1 + \frac{1}{2} + \frac{1}{3} + \cdots + \frac{1}{n} \tag{13.5}$$

Notice that H_n is a *partial sum*, the sum of the first n terms, of the harmonic series. Since H_{n-1} is the sum of all the offsets of T_n, we see that H_{n-1} is the distance by which the tower T_n overhangs its base.

We will not prove that for each natural number n, T_n gives the greatest possible overhang among all towers of n bricks; but we will prove the surprising fact that the maximal towers overhang arbitrarily far. More precisely, we will show that for any number B, however large, there exists a partial sum of the harmonic series that is larger than B. We express this fact by saying that the harmonic series *diverges to plus infinity* $(+\infty)$.

Proposition 13.7. *For any preassigned bound B, the overhang of tower T_n becomes larger than B by taking n sufficiently large. In other words, the sum (13.5) becomes larger than B if n is chosen large enough.*

Proof of Proposition 13.7. It is sufficient to show that by choosing n sufficiently large we can be sure that H_n is as large as we please. This may seem paradoxical because $1/n$ becomes arbitrarily small by taking n sufficiently large. Is it possible that a sum of such small quantities can become larger than, say, $1,000,000$? Yes, but as we will see, n must be very large indeed. Our plan is to construct a sum S_n, simpler to evaluate than H_n, whose terms are all less than or equal to the corresponding ones of H_n. Moreover, we can easily show that S_n exceeds any bound provided that n is suitably chosen. In the first line below we write H_{16}; in the second line we write S_{16} with corresponding terms arranged in columns; and in the third line we rewrite S_{16} with certain terms accumulated.

$$H_{16} = 1 + \tfrac{1}{2} + \tfrac{1}{3} + \tfrac{1}{4} + \tfrac{1}{5} + \tfrac{1}{6} + \tfrac{1}{7} + \tfrac{1}{8} + \tfrac{1}{9} + \tfrac{1}{10} + \tfrac{1}{11} + \tfrac{1}{12} + \tfrac{1}{13} + \tfrac{1}{14} + \tfrac{1}{15} + \tfrac{1}{16}$$

$$S_{16} = 1 + \tfrac{1}{2} + \tfrac{1}{4} + \tfrac{1}{4} + \tfrac{1}{8} + \tfrac{1}{8} + \tfrac{1}{8} + \tfrac{1}{8} + \tfrac{1}{16} + \tfrac{1}{16} + \tfrac{1}{16} + \tfrac{1}{16} + \tfrac{1}{16} + \tfrac{1}{16} + \tfrac{1}{16} + \tfrac{1}{16}$$

$$= 1 + \tfrac{1}{2} \quad + \tfrac{1}{2} \qquad\quad + \tfrac{1}{2} \qquad\qquad\qquad\qquad + \tfrac{1}{2}$$

Note that each term in the second sum S_{16} is less than or equal to the corresponding term in the first sum H_{16}, which is (13.5) with $n = 16$. The second sum S_{16} is easy to compute because it contains two terms $1/4$, four terms $1/8$, and eight terms

1/16. The third sum above shows these partial sums, all of which are equal to 1/2. Thus, we have shown that (13.5) with $n = 16$ is greater than

$$1 + \frac{1}{2} + \frac{1}{2} + \frac{1}{2} + \frac{1}{2} = 3$$

This argument can easily be extended to show that for $n = 2^m$, (13.5) is greater than $\leftarrow P$ Dangerous curve! $1 + m/2$. In other words, the following inequality is true:

$$1 + \frac{1}{2} + \frac{1}{3} + \cdots + \frac{1}{2^m} > 1 + \frac{m}{2} \qquad (13.6)$$

It follows that for any number B, however large, by summing suitably many terms of the harmonic series, that is, by making n in (13.5) sufficiently large, we can obtain a sum that is larger than B. And since (13.5) represents the distance that a tower of n bricks overhangs its base, we see that such a tower can overhang arbitrarily far from its base. □

It is easy to use inequality (13.6) to find a number large enough to ensure, at least in theory, an overhang of 100 feet. Since half a brick length is 3.75″, we find that 100 feet is 320 half-brick lengths. We have just seen that a tower of 2^m bricks can overhang by more than $1 + m/2$. It is sufficient to find a value for m such that

$$1 + \frac{m}{2} > 320$$

This relationship is equivalent to $m > 638$. In other words, by using the stacking method (13.4), we can be sure of an overhang of at least 100 feet if we take the number of bricks equal to 2^{638}. This number is much too large because the estimate (13.6) is quite rough, even though it was sufficient for the proof of Proposition 13.7. Using a more accurate estimate (that we will not discuss here) for (13.5), we obtain 1.6×10^{139} for the required number of bricks—in other words, 16 followed by 138 zeros. For comparison, the number of electrons in the universe is said to be approximately 10^{77}.

In the next chapter, we will see how Archimedes used the lever to demonstrate properties of the sphere.

Chapter 14

Two Pearls

I will summon up from the dust—where his measuring rod once traced its lines—an obscure little man ... Archimedes. When I was quæstor in Sicily [in 75 B.C., 137 years after the death of Archimedes] I managed to track down his grave. The Syracusans knew nothing about it, and indeed denied that any such thing existed. But there it was, completely surrounded and hidden by bushes of brambles and thorns. I remembered having heard of some simple lines of verse which had been inscribed on his tomb, referring to a sphere and cylinder modeled in stone on top of the grave. And so I took a good look round all the numerous tombs that stand beside the Agrigentine Gate. Finally I noted a little column just visible above the scrub: it was surmounted by a sphere and a cylinder. I immediately said to the Syracusans, some of whose leading citizens were with me at the time, that I believed this was the very object I had been looking for. Men were sent in with sickles to clear the site, and when a path to the monument had been opened we walked right up to it. And the verses were still visible, though approximately the second half of each line had been worn away.[1]

—CICERO (106–43 B.C.), *Tusculan Disputations*, Book 5

Cicero doesn't tell what the verses on Archimedes' tomb said—even the legible part—but, as a substitute, we discuss two of Archimedes' mathematical pearls. It is appropriate to refer to these two results of Archimedes—the volume and surface area of a sphere—as *pearls* because pearls are generally spherical. Furthermore, pearls are not as hard as some other gems. Mathematical gems are generally brilliant and diamond-hard—the hardness of terse, immutable logic. The softness in the present case comes from the fact that these are pearls of *heuristic* mathematics—the mathematics of meaning and discovery in contrast to the mathematics of rigorous proof. Nevertheless, these mathematical pearls proved more lasting than the stone column of Archimedes' tomb.

Archimedes, because he recognized that these arguments lack rigor, invented a method—we will not discuss it here—the *method of exhaustion* that anticipated the discovery of the calculus eighteen centuries later.

The first pearl makes use of Archimedes' *method of equilibrium*—the use of mechanics to discover geometric results. The second pearl is based on the figure that Cicero discovered on Archimedes' tomb—a sphere inscribed in a cylinder. It is possible that the illegible verses on his tomb related to this problem. Plutarch tells us that Archimedes requested that this figure be placed on his tomb; it seems that Archimedes esteemed this result among the highest of his achievements.

The Volume of a Sphere

When Archimedes attacked the problem of the volume of a sphere, he knew the solution of two simpler problems: the volume of a cylinder and the volume of a cone. His clever idea is to find the volume of the sphere by balancing it against some arrangement of the other two solid objects. The proportions and relative sizes of the three objects are shown in Figure 14.1. The method requires that the density of the cylinder be four times as great as the density of the other two objects.

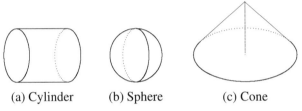

(a) Cylinder   (b) Sphere   (c) Cone

Figure 14.1. Objects to be balanced. The diameters of the cylinder and sphere are equal to the heights of both the cylinder and the cone. The radius of the cone is equal to the diameters of both the cylinder and sphere. The cone and sphere are equally dense, but the density of the cylinder is four times as great as the common density of the cone and the sphere.

The first step is to express the volumes of the cylinder and cone in terms of the area of a circle.

Fact 14.1. *The area of a circle of radius r is πr^2.*

Fact 14.2. *The volume of a cylinder is equal to the area of its base multiplied by its height. More precisely, a right circular cylinder with radius r and height h has volume $\pi r^2 h$.*

Fact 14.3. *The volume of a cone is equal to one-third the area of the base multiplied by its height. More precisely, a right circular cone with radius r and height h has volume $\frac{1}{3}\pi r^2 h$.*

With great ingenuity, Archimedes discovered the following formula for the volume of a sphere by showing that the volume of a sphere is expressible in terms of the volume of a certain cylinder and the volume of a certain cone.

Theorem 14.4. *A sphere of radius r has volume $\frac{4}{3}\pi r^3$.*

Archimedes uses Facts 14.2 and 14.3 to show that Theorem 14.4 is very plausible. However, we cannot call the following argument a proof because it is based on certain approximations.

Archimedes' name is popularly associated with the lever. Plutarch takes note of his use of the lever in designing catapults to protect the city of Syracuse during the siege by the Romans in 212 B.C.

> For Archimedes had provided and fixed most of his engines immediately under the wall; whence the Romans, seeing that infinite mischiefs overwhelmed them from no visible means, began to think they were fighting with the gods.
>
> —PLUTARCH, *Lives*

The engines of war were for nought because Syracuse fell, and Archimedes died in the ensuing disorder. But Archimedes' application of the lever to problems of mathematics was a success. The diagram in Figure 14.2(b) looks as though it might be one of Archimedes' siege engines. In fact, it pictures the equilibrium of a lever in which the cone and a sphere on the left side balance a cylinder of material four times as dense on the right side.

The relationship between the cone, sphere, and cylinder is shown in Figure 14.2(a). These three solid figures are formed by rotating three plane figures about the common axis FB. Thus, the square HIJK (with side $2r$ and center at C) generates the cylinder; the isosceles right triangle FLM (with right angle at F) generates the cone; and the circle FEBG (with radius r and center at C) generates the sphere. The cone, cylinder, and sphere intersect in the circle EG, a great circle with respect to the sphere. A vertical slice through all three solids is shown perpendicular to the common axis FB at an arbitrary distance x from C, the center of the sphere. We consider the line segment AFB to be a balance arm with fulcrum at F. In Figure 14.2(b), we move the cone and the sphere and hang them from point A while the cylinder remains in place. In the side view (the right half of Figure 14.2(a)) the large circle outlines the base of the cone, and the small circle outlines both the sphere and the cone.

We will see that in Figure 14.2(b) *the cone and sphere on the left side of the lever balance exactly against the cylinder on the right side.* (Note that we refer to the *solid* cone, sphere, and cylinder—not merely the surface.) In this diagram, the cylinder is in the same position as it was in Figure 14.2(a), but both the cone and the sphere have been rotated by 90° and are hung at A, the left end of the lever. Note that the cone and sphere have density 1, and the cylinder has density 4. The vertical slice of the cylinder and the horizontal slices of the cone and sphere are the same slices as those shown superimposed in Figure 14.2(a).

Why is the configuration in Figure 14.2(b) in equilibrium? To answer this question, we will make use of the theory of the lever. Let us examine the thin slices of the three solids that are shown superimposed in Figure 14.2(a) and separated in Figure 14.2(b). The left edge of the slices shown in Figure 14.2(a) is at an arbitrary distance x from the center of the sphere of radius r. Let Δx be the thickness of the slices, and let v_c, v_s, and v_k denote the volumes of the slice (RS in Figure 14.2(b)) of the cone, the slice (TU) of the sphere, and the slice (PQ) of the cylinder, respectively. (The Greek word for cylinder is *kýlindros*; hence, v_k.) Using Fact 14.2, we have

$$v_k = \pi r^2 \Delta x$$

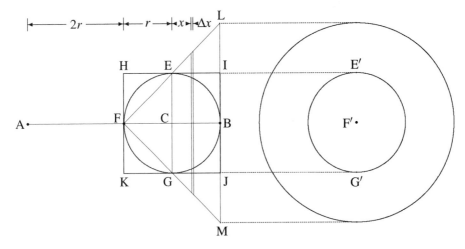

(a) Superimposed cylinder, sphere, and cone—front and side views. AB is the axis of the the cylinder and the sphere, and FB is a diameter of the sphere.

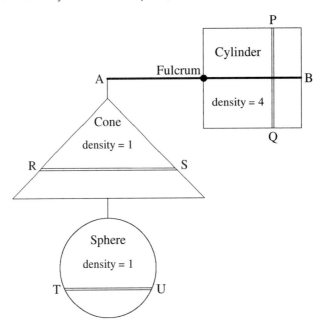

(b) Balancing the cone, cylinder, and sphere. As in (a), the axis of the cylinder (density = 4) is AB, but here the cone (density = 1) and sphere (density = 1) are suspended vertically from A. The system balances on the lever arm AB with respect to the fulcrum.

Figure 14.2.

Fact 14.2 cannot be used to give the *exact* volume v_c or v_s because these slices are not actually cylinders, but it gives a very good approximation if Δx is very small. Because it uses approximations, this argument becomes *heuristic* rather than rigorous. Indeed, Archimedes recognized this and supplemented this argument with a rigorous proof. In place of $=$ we use the sign $\approx$ for *approximately equal*.

In Figure 14.2(a) we see that the radius of the slice of the cone is $r+x$. In fact, this is the radius of the left face of the slice; we approximate by ignoring that the right face has a slightly larger radius. Using Fact 14.2, we obtain the following approximation for the volume of the slice of the cone:

$$v_c \approx \pi(r+x)^2 \Delta x = \pi(r^2 + 2xr + x^2)\Delta x$$

Now we must find the volume of the slice of the sphere. From the Pythagorean theorem the square of the radius of the slice of the sphere is $r^2 - x^2$. Now Fact 14.2 yields the approximation

$$v_s \approx \pi(r^2 - x^2)\Delta x$$

We suppose that the cylinder is more dense than the cone or the sphere. In particular, we assume that the cylinder has density equal to 4 (units of mass per unit volume) and that the cone and cylinder have density equal to 1. Now we study in Figure 14.2(b) the moments contributed by each of the slices with respect to the fulcrum.

Let let m_c, m_s, and m_k denote the masses of v_c, v_s, and v_k, respectively, and, similarly, let M_c, M_s, and M_k denote the respective moments. Using the assumed densities, we have

$$m_c = v_c \quad m_s = v_s \quad m_k = 4v_k$$

Recall that in computing moments we may concentrate the mass of each slice to a point mass located at the center of mass. Horizontal distances from the fulcrum are measured positive to the right and negative to the left. Let d_c, d_s, and d_k be the respective distances of these point masses from the fulcrum. Since the point masses representing the slices RS (v_c) and TU (v_s) are located directly below point A, the end of the lever arm, the horizontal distance of both of these point masses is

$$d_c = d_s = -\overline{\text{AF}} = -2r$$

On the other hand, $d_k = r+x$. The total moment is

$$
\begin{aligned}
d_c m_c &+ d_s m_s + d_k r_k \\
&= -2rv_c - 2rv_s + (r+x)(4v_k) \\
&\approx -2r\pi(r^2 + 2xr + x^2)\Delta x - 2r\pi(r^2 - x^2)\Delta x + 4(r+x)\pi r^2 \Delta x \\
&= \pi\{(-2r^3 - 4xr^2 - 2x^2 r) + (-2r^3 + 2x^2 r) + (4r^3 + 4xr^2)\}\Delta x \\
&= 0
\end{aligned}
$$

Thus, we see that the slices balance. But in Figure 14.2(a) we can cut the cone, sphere, and cylinder into a very large number of thin slices. Each triple of slices balances. This means that the entire configuration in Figure 14.2(b) balances. It can be shown using calculus that our approximation is good enough so that Figure 14.2(b) is precisely—not just approximately—in balance.

How can we use this fact to find the volume of a sphere of radius r? In Figure 14.2(b), let V_c, V_s, and V_k denote the total volumes of the cone, sphere, and cylinder, respectively. From Facts 14.2 and 14.3, we have

$$V_k = \pi r^2(2r) = 2\pi r^3$$
$$V_c = \frac{1}{3}\pi(2r)^2(2r) = \frac{8}{3}\pi r^3$$

In computing moments we can consider these figures to be point masses concentrated at their centers of mass. Since the center of mass of the cylinder is at distance r to the right of the fulcrum, the moment contributed by the cylinder of density 4 is equal to

$$4rV_k = 4r \cdot 2\pi r^3 = 8\pi r^4$$

The sum of the (unsigned) moments of the cone and sphere of density 1 must be equal to moment of the cylinder.

$$2r \cdot \frac{8}{3}\pi r^3 + 2rV_s = 8\pi r^4$$

First dividing by $2r$, we solve this equation for V_s.

$$\frac{8}{3}\pi r^3 + V_s = 4\pi r^3$$
$$V_s = 4\pi r^3 - \frac{8}{3}\pi r^3 = \frac{4}{3}\pi r^3$$

This concludes a very strong plausibility argument for Theorem 14.4.

The Surface Area of a Sphere

To find the area of a sphere, we must first know the surface area of two simpler figures—the cylinder and the cone. These surfaces are simpler than the sphere because they can be formed by bending a plane region. A surface with this property is said to be *developable*. The lateral surface of a cone or a cylinder is developable, but the surface of a sphere is not. To measure the surface area of a developable surface, first peel the surface—like the label of a tin can—and flatten it. In the case of a cylinder, the result is a rectangle, and we find the area by multiplying the width and height.

Proposition 14.5. *The lateral surface area of a right circular cylinder of radius r and height h is equal to the perimeter multiplied by the height, that is, $2\pi rh$.*

The result of flattening the lateral surface of a cone is a sector of a circle. Thus, the problem of the surface area of a cone reduces to finding the area of a sector of a circle.

Proposition 14.6. *The area of a circular sector of arc length S and radius R is RS/2.*

Proof. The area of the sector is what fraction of the area of an entire circle? The ratio of the areas is the same as the ratio of the arc length of the sector to the arc length of the entire circle, that is, the ratio $S : 2\pi R$. Putting A for the area of the sector, we have

$$\frac{A}{\pi R^2} = \frac{S}{2\pi R}$$

Solving for A we have $A = RS/2$. □

It is interesting to compare this result with the formula for the area of a triangle: $bh/2$ where b is the length of the base and h is the height of the triangle.

Proposition 14.7. *The lateral surface area of a right circular cone of radius r and slant height s is equal to the circumference of the base multiplied by the slant height, that is, πrs.*

Proof. In Figure 14.3, the arc length of B′C′B″ is equal to the circumference of the base of the cone; that is, $2\pi r$. Hence, according to Proposition 14.6, the area of the sector A′B′C′B″ is equal to πrs. □

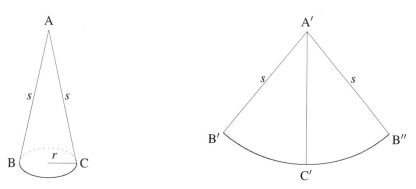

Figure 14.3. The lateral surface area of cone ABC. The base BC of the cone is a circle of radius r. The slant height AB of the cone is equal to s. The result of peeling the surface and flattening it is a sector of a circle of radius s, A′B′C′B″. The cone is cut along its left edge AB which appears in the sector as the lines A′B′ and A′B″. The line A′C′ is the image of the right edge AC of the cone.

Definition 14.8. The portion of a cone contained between two planes perpendicular to the axis of the cone is called a *frustum* of a cone.

Using Proposition 14.7, it is straightforward to find the surface area of a frustum of a cone, but we will omit the proof.

Fact 14.9. *Let $\mathcal{F}$ be a frustum of a cone with slant height s, and r_1 and r_2 be the radii of the bottom and top, respectively. The lateral area of $\mathcal{F}$ is equal to the slant height multiplied by the average perimeter—the average of top and bottom. That is, the area of $\mathcal{F}$ is equal to $\pi(r_1 + r_2)s/2$.*

We are now ready to give Archimedes' argument for the plausibility of the following assertion.

Theorem 14.10. *The surface area of a sphere of radius r is* $4\pi r^2$.

Referring to Figure 14.4, the slice PQ of the cylindrical surface is a thin cylinder whose area, according to Proposition 14.5, is equal to $2\pi r\Delta x$.

On the other hand, we *approximate* the area of the slice VW of the spherical surface by using the formula for the area of a frustum of a cone. Although it is not actually a frustum of a cone, we obtain a very close approximation to the surface area in this way, especially if the slice is very thin. We define the approximating frustum as follows. Starting in Figure 14.4 from the point V, which is on the circle of radius r, draw a *straight line* VX perpendicular to the radius CV and terminating at the right edge of the slice. Point X is very nearly on the circle. We use the length $\overline{VX}$ as an approximation for the circular arc. By rotating the segment $\overline{VX}$ about the axis AB, we obtain a very thin frustum of a cone. We use the area of this frustum as an approximation for the spherical area within the slice. To compute the surface area of the frustum, we need its average circumference and slant height.

Instead of finding the average circumference of the frustum, a rougher approximation is sufficient for our purpose. We use instead the perimeter of the left side of the slice, the cut that is at distance x from C, the center of the sphere. Using the Pythagorean theorem once again, the radius $\overline{ZV}$ of this cut through the sphere is $\sqrt{r^2 - x^2}$; therefore, the circumference is equal to $2\pi\sqrt{r^2 - x^2}$.

Slant height. The tiny triangle XYV is similar to triangle CZV by Fact 12.6 on page 176, because corresponding sides are perpendicular. Therefore, by equation (12.1) on page 175, we have

$$\overline{VX}/\overline{VY} = \overline{CV}/\overline{ZV}$$

As previously remarked, by the Pythagorean theorem we have

$$\overline{ZV} = \sqrt{r^2 - x^2}$$

Since $\overline{CV} = r$ and $\overline{VY} = \Delta x$, the previous equality leads to

$$\overline{VX} = \frac{r}{\sqrt{r^2 - x^2}}\Delta x$$

Now we can compute our approximation for the surface area of the frustum.

$$\begin{aligned}
\text{Area} &= \text{Perimeter} \times \text{Slant height} \\
&= 2\pi\sqrt{r^2 - x^2} \cdot \frac{r}{\sqrt{r^2 - x^2}}\Delta x \\
&= 2\pi r\Delta x
\end{aligned}$$

Thus, we have found that the thin cylindrical slice PQ and the thin spherical slice VW have approximately the same surface area. The approximation becomes better as the slices become thinner. By adding a large number of very thin slices, it is

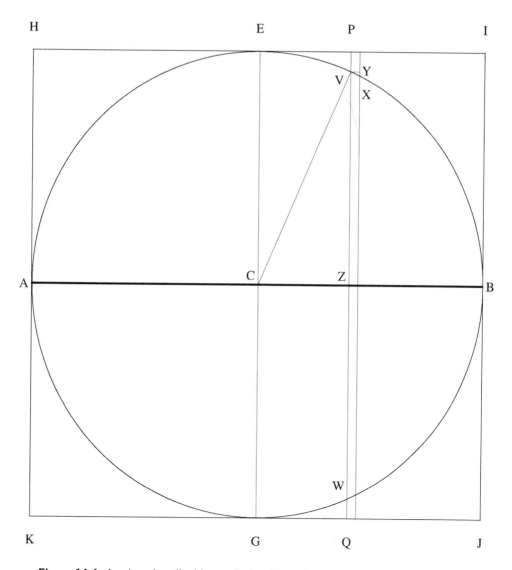

Figure 14.4. A sphere inscribed in a cylinder. The cylinder is generated by rotating the square (of side $2r$) HIJK about the axis AB, and the sphere is generated by rotating the circle (of radius r) AEBG similarly about AB. This figure is like Figure 14.2(a) without the cone, except that here we are concerned only with the surfaces: the surface of the sphere and the lateral surface of the cylinder. As in the previous figure, we make thin vertical slices (of width Δx) simultaneously: slice VW of the surface of the sphere and slice PQ of the lateral surface of the cylinder. We show that the area of slice VW is approximately equal to the area of slice PQ. In fact, these areas are *exactly* equal, even if the slices are not thin. Since the slices could be the entire sphere and the entire cylinder, it follows that the sphere and the cylinder have the same surface area.

plausible that the entire surface area of the sphere is equal to the entire surface area of the cylinder which, by Proposition 14.7, is equal to

$$2\pi r \cdot 2r = 4\pi r^2$$

> *But nothing afflicted Marcellus [the Roman general] so much as the death of Archimedes; who was then, as fate would have it, intent upon working out some problem by a diagram, and having fixed his mind alike and his eyes upon the subject of his speculation, he never noticed the incursion of the Romans, nor that the city was taken. In this transport of study and contemplation, a soldier, unexpectedly coming up to him, commanded him to follow to Marcellus; which he declining to do before he had worked out his problem to a demonstration, the soldier, enraged, drew his sword and ran him through. Others write, that a Roman soldier, running upon him with a drawn sword, offered to kill him; and that Archimedes, looking back, earnestly besought him to hold his hand a little while, that he might not leave what he was then at work upon inconclusive and imperfect; but the soldier, nothing moved by his entreaty, instantly killed him. Others again relate, that as Archimedes was carrying to Marcellus mathematical instruments, dials, spheres, and angles, by which the magnitude of the sun might be measured to the sight, some soldiers seeing him, and thinking that he carried gold in a vessel, slew him. Certain it is, that his death was very afflicting to Marcellus; and that Marcellus ever after regarded him that killed him as a murderer; and that he sought for his kindred and honored them with signal favors.*
>
> *—PLUTARCH, Lives*

Part IV

NUMBERING

Chapter 15

New Numbers for Old

Minus times minus equals plus,
The reason for this we need not discuss.

—ANONYMOUS

The discussion of *minus times minus* leads to questions concerning the nature of various extensions of the natural numbers. We will examine numbers from the point of view of mathematical models and axiomatic systems, and we will discuss the construction of the integers from the natural numbers and the complex numbers from the real numbers. We will also see that the complex numbers are rich with geometric meaning.

Minus Times Minus Equals Plus

Negative numbers are part of everyday life. The words *credit* and *debit* may be more familiar, but the meaning is still negative and positive. Even more familiar are the words *before* and *after*. However, the definition of B.C. and A.D. is an example that shows a curious inconsistency with the number system. Since there is no year zero—an omission due to the monk Dionysius Exiguus (Denis the Little) in the sixth century—periods of time from B.C. to A.D. are one year shorter than might appear. For example, Archimedes was born in 287 B.C.; if it is now A.D. 1999, how many years ago was that? Although $1999 - (-287) = 2286$, he was actually born 2285 years ago.

This practical familiarity with negative numbers does not provide an answer to the question, Why is it true that $(-1) \times (-1)$ is equal to $+1$? Even the great mathematician Leonhard Euler is said to have talked nonsense on this point. He is reported to have said that $(-1) \times (-1)$ cannot be equal to -1 because $(+1) \times (-1)$ is equal to -1, and therefore—perhaps by process of elimination—$(-1) \times (-1)$ must equal $+1$. Below we will see more convincing arguments.

The first step in extending the natural numbers is to define the negative integers and zero. We must do so in a way that meets the needs of accountants and others. We also want to extend the system to be an *arithmetic*,[1] and we want the extended system to be free of contradictions. We will discuss the details of this extension later, but for now we will base our demonstration concerning $(-1) \times (-1)$ on the following facts.

$$1 + (-1) = 0 \tag{15.1}$$
$$(-1) \times 0 = 0 \tag{15.2}$$
$$(-1) \times (+1) = -1 \tag{15.3}$$

In addition, we use the fact that the extended system is an arithmetic. In particular, we make use of the *distributive law*, which states that for arbitrary numbers a, b, and c the following equality holds:

$$a(b+c) = ab + ac$$

Our task is to show that equations (15.1), (15.2), (15.3), and the distributive law imply

$$(-1) \times (-1) = +1 \tag{15.4}$$

Proposition 15.1. *Equation* (15.4) *is correct.*

Proof. Apply the distributive law with $a = -1$, $b = 1$, and $c = -1$. We obtain

$$(-1) \times (1 + (-1)) = (-1) \times (1) + (-1) \times (-1) \tag{15.5}$$

The left side of equation (15.5) is equal to 0 because of (15.1) and (15.2), and the right side is equal to $-1 + (-1) \times (-1)$ because of (15.3). Now equation (15.4) follows from adding 1 to both sides of the following equation:

$$-1 + (-1) \times (-1) = 0 \qquad\qquad \square$$

Equation (15.4) has a geometric interpretation. Visualize a number line with the integers marching in increasing order from left to right:

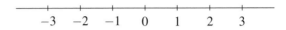

The integers are a *mathematical model* for a system of equally spaced points on a line, and the numerals $-3, -2, \ldots$ are labels to tell us what number is represented by a particular point.

Now consider the effect of transforming this line by multiplying every number by -1. Using *minus times minus equals plus*, the result is the same number line, except that now the integers march in increasing order from right to left.

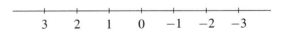

We see that multiplication by -1 transforms the number line exactly as would a rotation by $180°$ in the plane of this sheet of paper about the point zero. For example, under this rotation, the points 1, 2, and 3 exchange places with the points -1, -2, and -3. (Of course, we ignore the fact that the numerals are turned upside-down in this process.) We see that the geometric rotation by $180°$ corresponds to the arithmetic multiplication by -1. With this geometric interpretation, equation (15.4) loses its mystery. Indeed, it is more comprehensible to consider numbers as a mathematical model for a sequence of points on a line than as an axiomatic system that satisfies, among other properties, the arcane distributive law.

We have considered two modes of understanding numbers: (1) numbers as a mathematical model and (2) numbers as an axiomatic system. However, there is another mode: (3) the *construction* of the number system, in which more complicated number concepts are built up from simpler concepts.

Three Views of the Number System

Let us spend a few minutes discussing these three modes of understanding the number system in order of difficulty.

1. *Numbers as a mathematical model:* What are the interpretations and uses of the numbers? By a mathematical model we mean the use of mathematical concepts to represent something in the world outside of mathematics. Our discussion of rotation is one example. Here are a few among many other examples.

   (a) The natural numbers model our experience of counting objects. This is without doubt the most fundamental and important of all mathematical models.

   (b) By augmenting the natural numbers with zero and the negative numbers, we obtain the system called the *integers*. The integers model, among other things, the keeping of accounts.

   (c) The rational numbers describe the division of physical objects—for example, the division of land.

   (d) The real numbers model our experience of making linear measurements.

2. *Numbers as an axiomatic system:* How do the numbers *work*? What rules govern them? This thinking underlies our first discussion—the one involving the distributive law. Instead of trying to understand what numbers *are*, we set down certain properties—we call them *axioms*—that are demonstrably true for the natural numbers, and we require that any extension of the number system beyond the natural numbers should also satisfy these axioms. Here is a more complete list of axioms. Let a, b, and c denote arbitrary numbers—they could

be natural numbers, or numbers from some extension of the number system such as the integers or the rational numbers.

$$\text{The commutative law of addition:} \quad a+b=b+a \quad (15.6)$$
$$\text{The associative law of addition:} \quad a+(b+c)=(a+b)+c \quad (15.7)$$
$$\text{The commutative law of multiplication:} \quad ab=ba \quad (15.8)$$
$$\text{The associative law of multiplication:} \quad a(bc)=(ab)c \quad (15.9)$$
$$\text{The distributive law:} \quad a(b+c)=ab+ac \quad (15.10)$$

Mathematics requires confidence in a chain of logical argument. Logic plays a more dominant role in mathematics than in ordinary life because we have more confidence in the premises of mathematical argument—in the axioms—than we have in the premises that underlie logical arguments in ordinary life.

However, the axiomatic method does have a flaw. How do we know that an axiomatic system is consistent? We would be convinced if we could see that there are objects that actually exist that satisfy the axioms. *Existence* dispels the suspicion of inconsistency. The question of existence is another way of asking the question, *What are the numbers?* and this leads to the third mode of understanding the number system.

3. *The constructive representation of the number system.* This process can start with the acceptance of the natural numbers as undefined entities. In the words of the mathematician Leopold Kronecker (1823–91), "God created the natural numbers; everything else is the work of humanity." The *construction* of the number system is a difficult task. Moreover, some may find it difficult to believe that the various kinds of numbers really *are* such complicated structures. However, we can get the flavor of this process by developing two of these constructions, the integers and the complex numbers.

The Integers

In this section we will show how the integers (including the negative integers and zero) can be built out of the natural numbers. In fact, we will *define* the integers in terms of the natural numbers. Our reason for extending the natural numbers is to make it always possible to subtract one number from another. For example, from the perspective of the natural numbers we cannot subtract 5 from 4 because there is no natural number -1. (Strictly speaking, we should not even talk about -1 until we have finished this construction.)

Imagine a *Mad Subtractor* who tries unsuccessfully again and again to subtract pairs of natural numbers. Mad Subtractor has a desk with infinitely many pigeonholes. Whenever he fails at one of his subtraction problems, he obsessively files the troublesome pair in one of the pigeonholes. But his filing is not willy-nilly because he files together those subtraction pairs that he feels should have the same answer. For example, he puts $(4,5)$ together with $(7,8)$ and $(12,13)$. In his despair, he decides to file *all* pairs of natural numbers in this way, even if the difference is actually a natural number. Just as he finishes his filing task, he comes to a remarkable epiphany. He exclaims, "The pigeonholes themselves are the numbers that I am seeking. Yes, I will call these pigeonholes *integers*. Then I can subtract natural numbers whenever ↪*P* Dangerous curve! I wish, and the result will always be an integer, the integer/pigeonhole that contains, among others, the pair I wish to subtract."

Bizarre though it might seem, we claim that there is merit in the idea that each integer *is* a collection of pairs of natural numbers. Let us supply some details to this vision of the Mad Subtractor. We begin by considering ordered pairs (a, b) of natural numbers. We make the following definition that we will use only in this section.

Definition 15.2. Let (a,b) and (c,d) be ordered pairs of natural numbers. We say that these two pairs are *equivalent* and we write

$$(a,b) \sim (c,d) \qquad \text{if} \qquad a+d = b+c$$

In the above definition, we would like to require $a - b = c - d$, but we cannot do so because, as the Mad Subtractor discovered, this requirement is not always meaningful. However, it is meaningful to require the equivalent condition, $a + d = b + c$.

It can be shown that all the ordered pairs of natural numbers can be divided into nonoverlapping classes, called equivalence classes,[2] such that two pairs belong to the same class if and only if they are equivalent. Of course, the equivalence classes are the pigeonholes of the Mad Subtractor.

We define addition and multiplication of ordered pairs of natural numbers as follows:

$$(a,b) + (c,d) = (a+c, b+d) \qquad (a,b) \cdot (c,d) = (ac+bd, bc+ad)$$

The motivation for these definitions is that we expect the following relations to be true for the integers.

$$(a-b) + (c-d) = (a+c) - (b+d)$$
$$(a-b)(c-d) = (ac+bd) - (bc+ad)$$

It can be shown that the definitions of addition and multiplication of ordered pairs satisfy the commutative, associative, and distributive laws. Moreover, it can be shown that the *equivalence class that contains a sum (or product) depends only on the equivalence classes that contain the two summands (or factors, respectively)*. This means that we have defined an addition and multiplication of equivalence classes by means of arbitrary *representatives* of the classes.

The equivalence classes are called the *integers*—positive, zero, and negative. These are the same integers that are familiar to us. What is different is that now we know what the integers really *are*.

Let (a,b) belong to a certain equivalence class. There are three possibilities:

1. $a > b$. The equivalence class is denoted n where $n = a - b$. Thus, we use exactly the same notation for the integer 5 that we use for the natural number 5

without the risk of confusion because the arithmetic of the positive integers is exactly the same as the arithmetic of the natural numbers. When we calculate, we never need to know whether we are dealing with a natural number or an equivalence class of natural numbers.

2. $a = b$. The equivalence class is denoted 0.

3. $a < b$. The equivalence class is denoted $-n$ where $n = b - a$.

> Q. Before we started this exercise, we knew how to compute with the integers. What do we know now about the integers that we didn't know before?
>
> A. We now know that the integers, since they are constructed from the natural numbers, are free of contradictions provided that the natural numbers are free of contradictions.

This development of the integers provides another answer, indeed the conclusive answer, to the question raised on page 15. We can actually carry out the computation $(-1) \times (-1)$. In fact, the integer -1 is the equivalence class containing, for example, $(1, 2)$. We apply the definition of multiplication of ordered pairs to obtain

$$(1, 2) \cdot (1, 2) = (1 \cdot 1 + 2 \cdot 2, 2 \cdot 1 + 1 \cdot 2) = (5, 4)$$

However, since $5 - 4 = 1$, the pair $(5, 4)$ belongs to the equivalence class that we denote $+1$. Thus, we have shown $(-1) \times (-1) = +1$.

The Complex Numbers

There is a method of building the real numbers starting with the natural numbers, but we will not do this. We begin by accepting the real numbers, the *infinite decimals*, as known. We define an extension of the real numbers called *complex numbers*.

Complex numbers were introduced in the sixteenth century by Italian mathematicians[3] to solve certain algebraic equations. They introduced complex numbers as a means of solving problems with no apparent connection to complex numbers. For example, Cardano's method for solving cubic equations uses complex numbers in the solution of certain equations, even though neither the equation nor the numbers that satisfy the equation—the solutions—are complex. Although the method bears Cardano's name, credit for its discovery was the subject of heated contention between Cardano, Tartaglia, and Ferrari; it was probably discovered by del Ferro.

The simplest equation that requires complex numbers for its solution is

$$x^2 = -1 \tag{15.11}$$

In fact, it is clear that no real number satisfies equation (15.11) because the square of a real number must be nonnegative. We define a new system of numbers, the complex numbers, in which this equation does have a solution. We define the complex numbers to consist of ordered pairs (a, b) of real numbers. We intend that the pair

$(0,1)$, which we denote i, will be a solution of equation (15.11). In addition, we want pairs of the form $(a,0)$ to be just another notation for the real numbers. In fact, we use a to denote the ordered pair $(a,0)$. When we write the number 6, it makes no difference in computations whether we intend the natural number 6, the integer 6, the rational number 6, the real number 6, or the complex number 6. Since i is a solution of equation (15.11), we must have $i^2 = -1$; that is,

$$(0,1) \cdot (0,1) = (-1,0) \tag{15.12}$$

The addition of complex numbers and multiplication of a real number by a complex number are defined

$$(a,b) + (c,d) = (a+c, b+d)$$
$$(a,0) \cdot (c,d) = (ac, ad)$$

for arbitrary a, b, c, and d. As a consequence of these definitions, we can represent the ordered pair (a,b) as $a + ib$.

We require that the associative, commutative, and distributive laws remain true for the complex numbers. This forces us to define the multiplication of arbitrary complex numbers as follows.

$$\begin{aligned}(a,b) \cdot (c,d) &= (a+ib)(c+id) = ac + i(bc+ad) + i^2 bd \\ &= ac + i(bc+ad) - bd = (ac - bd, bc + ad)\end{aligned} \tag{15.13}$$

We summarize the addition and multiplication of complex numbers as follows.

Definition 15.3. Let (a,b) and (c,d) be arbitrary complex numbers. We define addition and multiplication as follows:

$$(a,b) + (c,d) = (a+c, b+d) \tag{15.14}$$
$$(a,b) \cdot (c,d) = (ac - bd, ad + bc). \tag{15.15}$$

Using this definition, it is possible to show that the axioms (15.6) through (15.10) are all true. The calculations needed to show this are tedious but perfectly straightforward.

Definition 15.4. Let $c = (a,b)$ be an arbitrary complex number. Then a and b are called, respectively, the real and imaginary part of c. The real and imaginary parts of c are often denoted $\Re(c)$ and $\Im(c)$, respectively.

Notation. The symbols $\Re$ and $\Im$ that are used for the real and imaginary parts of complex numbers are the letters R and I in the German Fraktur font. In handwritten material, ordinary cursive letters may be used.

The letter i stands for *imaginary*, but there is nothing imaginary in the ordinary sense about the number i. This is an unfortunate term that was applied centuries ago when these numbers were new and not well understood.

Q. You surely don't mean that the complex numbers are as meaningful as the real numbers that we know and love, do you?

A. Yes, that is exactly what I mean. The complex numbers have just as much actual existence as the real numbers. Complex numbers are used extensively in physical science, a study firmly rooted in the real world. Both the complex and the real numbers are formidable intellectual constructions, but both systems are equally meaningful and reliable. Complex numbers can be used with the same security and confidence with which we use real numbers.

Q. Defining the complex numbers seems like a game. Is it possible to continue defining more and more different kinds of numbers?

A. This process of constructing the complex numbers does seem like a game, but that impression changes when we see the richness of mathematical models based on the complex numbers. The complex numbers have a unique place in science. Although there are further extensions that are called hypercomplex numbers,[4] they are of much less importance than the complex numbers. This is true in part because these extensions cannot preserve all of the laws: associative, commutative, and distributive.

We are comfortable with the real numbers because they generalize our everyday experience of making linear measurements. To become more comfortable with the complex numbers we need a similar interpretation. In fact, the complex numbers are a model for a two-dimensional plane (Figure 15.1)—the *complex plane*. We are familiar with the method of presenting data consisting of ordered pairs in the form of a two-dimensional chart. Every day, the financial page of the newspaper uses charts of this sort—for example, showing the prices of particular stocks through a period of time. Such a chart is possible because each (day, price) pair corresponds uniquely with a point in the plane.

To represent the complex numbers as points in a plane we fix arbitrarily two perpendicular lines (see Figure 15.1) that intersect in a point called the origin. Both lines carry copies of the real numbers—on the horizontal number line, called the real

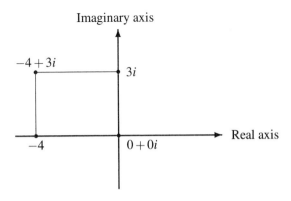

Figure 15.1. The complex plane showing the location of $-4+3i$.

axis, the numbers increase to the right and on the vertical number line, called the imaginary axis, the numbers increase upward.

There are geometric interpretations of both addition and multiplication of complex numbers. For addition, the interpretation is straightforward. The sum of two complex numbers is obtained by a process of completing a parallelogram (see Figure 15.2).

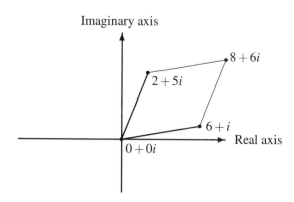

Figure 15.2. Addition of complex numbers. The sum $(6+i) + (2+5i) = 8+6i$ is found by completing a parallelogram starting with the two sides, shown here as the thick lines that meet at the origin, $0+0i$, and that terminate in the complex numbers to be added, $6+i$ and $2+5i$. The sides, shown here as thin lines, that complete the parallelogram meet at $8+6i$, the sum of $6+i$ and $2+5i$.

Multiplication of Complex Numbers

It is more difficult to give a geometric meaning to complex multiplication. To do this, we must introduce a couple of definitions.

Definition 15.5. Let $z = x+yi$ be a complex number. The *modulus* of z, denoted $|z|$, is defined to be $\sqrt{x^2+y^2}$. (See Figure 15.3.)

> In number theory, the word *modulus* has a different meaning. See page 93.
>
> The Greek letter *theta* (θ) is often used to represent angles.
>
> **Pronunciation.**
> (U.S.) *THAY·ta*
> (U.K.) *THEE·ta.*
> (*TH* as in *thin.*)

It is a consequence of the Pythagorean theorem that $|z|$ is just the distance between z and the origin $0+0i$. The set of all complex numbers of modulus r, where r is an arbitrary positive number, consists of a circle in the complex plane of radius r with its center at the origin.

The following fact states two important properties concerning the modulus of complex numbers. We omit the proofs, which are straightforward applications of Definition 15.5.

Fact 15.6. *Let z_1 and z_2 be arbitrary complex numbers.*

1. *The distance between z_1 and z_2 is equal to $|z_1 - z_2|$.*

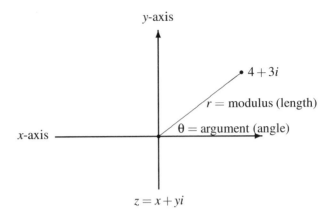

y-axis

x-axis

$z = x + yi$

Figure 15.3. Modulus and argument of the complex number $4+3i$. The modulus of $4+3i$ is the length r of the line segment from the origin to $4+3i$. By the Pythagorean theorem, the value of the modulus $r = |4+3i|$ is 5. The argument θ of $4+3i$ is the angle between the positive x-axis and the same line segment.

2. *The modulus of the product $z_1 z_2$ is the product of the moduli, that is,*

$$|z_1 z_2| = |z_1||z_2|$$

The following definition is a companion of Definition 15.5.

Definition 15.7. Let z be a nonzero complex number. The angle formed by the positive real axis and the line segment from $0 + 0i$ to z is called the *argument* of z. (See Figure 15.3.)

To avoid the ambiguity that, for example, $180°$ represents the same angle as $-180°$, we usually require that the argument of a complex number is nonnegative and less than $360°$.

The unique complex number of modulus 1 and argument θ is denoted $\operatorname{cis}\theta$. Here are a three examples:

$$\operatorname{cis}0° = 1 \qquad \operatorname{cis}180° = -1 \qquad \operatorname{cis}90° = i$$

A complex number is completely determined by its modulus and argument. In fact, the unique complex number with modulus r and argument θ is $r\operatorname{cis}\theta$. This is called the *polar representation* of a complex number.

The real and imaginary parts of $\operatorname{cis}\theta$ have special names. The real part of $\operatorname{cis}\theta$ is call the *cosine* of θ and is written $\cos\theta$, and the imaginary part is called the *sine* of θ and is written $\sin\theta$, so that, by definition,

$$\operatorname{cis}\theta = \cos\theta + i\sin\theta$$

In fact, the term cis is an acronym for <u>c</u>os + <u>i</u> <u>s</u>in.

> Q. *Don't sines and cosines have to do with trigonometry?*
>
> A. *Yes, these are the sines and cosines of trigonometry. The above definition leads to derivations of all the important facts of trigonometry, but let us return to complex multiplication.*

Mapping the Complex Plane

The mapping *w=cz*

Consider the product of two complex numbers, $c = a + bi$ and $z = x + yi$. Here we are following a common practice in mathematics of using the letters at the beginning of the alphabet to represent constants and letters at the end of the alphabet to represent variables.

Now look at what happens to the product cz when we keep c fixed and allow z to have all possible complex values.

As you recall from our previous discussion of functions, when one quantity *depends* on the value of another quantity, we say that the one is a function of the other. That is the situation here, although the quantities are now complex. The study of functions of a complex variable is an enormous branch of mathematics that is still the subject of current research—hundreds of articles are published every year. There are advanced courses devoted to this subject at all universities. This field of study has important applications in technology, for example, in the design of airplane wings.

Functions are often called *mappings*. Let us define a mapping from the z-plane to the w-plane by means of the relationship $w = cz$ where c is a constant. The term *mapping* is particularly appropriate here because in ordinary usage a map is two-dimensional and the complex plane is also two-dimensional. The connection may seem tenuous now, but when we have seen a few examples, this usage of the word *mapping* will seem very natural. We will study the mapping properties of $w = cz$ for a few special choices of the constant c.

Example 15.8. Put c equal to 2. What does it mean geometrically to multiply every number in the complex z-plane by 2?

Solution. The equation $w = 2z$ defines a mapping, also called a transformation, of the complex z-plane to the complex w-plane—each point in the z-plane has an image point in the w-plane. Putting $z = x + yi$ and $w = u + vi$, the equation $w = 2z$ becomes $u + vi = 2x + 2yi$, and, since complex numbers are equal only if their real and imaginary parts are equal, this can also be written

$$u = 2x, \ v = 2y$$

The effect of the mapping is to place the image point in the w-plane on the same ray from the origin as the object point in the z-plane but at double the distance from the origin. (See Figure 15.4.) Multiplication by an arbitrary positive number has a similar meaning. A mapping of this sort that simply multiplies the distance of each point from the origin by the same positive quantity is called a *similarity transformation*.

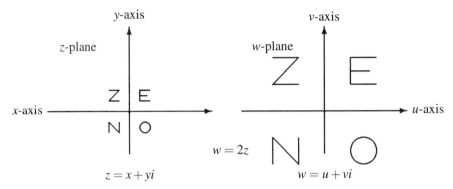

Figure 15.4. The mapping $w = 2z$. Each image point w is on the same ray through the origin as z at double the distance to the origin. A mapping of this sort that stretches (or diminishes) the distance to the origin equally in all directions is called a *similarity transformation*. Replacing 2 by a positive constant c in the above mapping, we obtain a general similarity transformation $w = cz$.

The similarity transformations are precisely the mappings $w = cz$ where c is real and positive.

Example 15.9. Suppose that c is equal to -1.

Solution. A short while ago we found that multiplying the *integers* by -1 is the same as rotating the number line about zero by 180°. Surprisingly, we see here that the same interpretation holds in the complex plane. (See Figure 15.5.) Multiplication by -1 maps the complex plane by rotating it about the origin by 180°.

Example 15.10. Put c equal to i.

Solution. What does it mean geometrically to multiply an arbitrary complex number by i? In other words, what is the geometric meaning of the mapping $w = iz$?

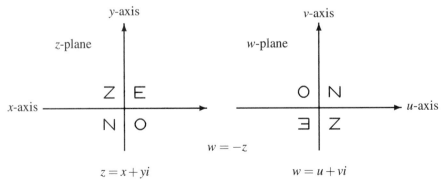

Figure 15.5. The mapping $w = -z$. The letters ZENO in the z-plane become rotated by 180° in the w-plane.

Recalling that $i^2 = -1$, we have

$$u + vi = i(x + yi) = ix + yi^2 = -y + xi$$

But, as we have seen, if two complex numbers are equal, it means that both their real and their imaginary parts are equal. In other words, we have

$$u = -y, \ v = x$$

As we see in Figure 15.6, the geometric effect of the mapping is rotation counterclockwise by 90°.

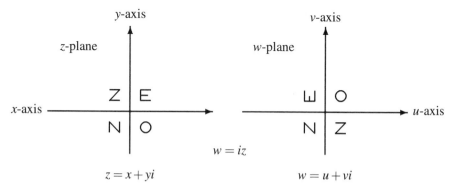

Figure 15.6. The mapping $w = iz$ from the z-plane to the w-plane means rotation counterclockwise about the origin by 90°. In terms of x, y, u, and v, the mapping is given by $u = -y$, $v = x$.

> Q. Is multiplication by a complex number always equivalent to a suitable rotation about the origin?
>
> A. Multiplication by a complex number of unit length is equivalent to a pure rotation. Don't forget that multiplication by a positive number corresponds not to a rotation but to a similarity transformation.

We have considered the mapping properties of $w = cz$ for two cases in which $|c| = 1$, namely, the cases $c = -1$ and $c = i$. Now we consider the general case, the case in which $c = \operatorname{cis} \theta$ for an arbitrary angle θ.

Proposition 15.11. *Let θ be an arbitrary angle, and let z be an arbitrary complex number. Counterclockwise rotation of z by the angle θ about the origin locates the point $z \operatorname{cis} \theta$.*

In other words, under the mapping $w = z \operatorname{cis} \theta$ the image points in the w-plane are located by rotating the z-plane counterclockwise about the origin by the angle θ as shown in Figure 15.7.

Proof. The proof consists of four steps.

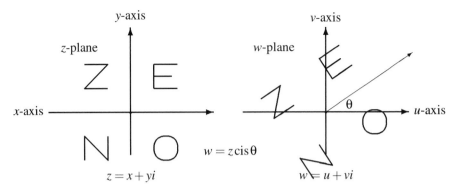

Figure 15.7. The mapping $w = z\operatorname{cis}\theta$ from the z-plane to the w-plane where θ is the angle labeled in the w-plane. The mapping has the geometric effect of rotation counterclockwise by the angle θ.

1. *We observe that the origin is mapped into itself.*

2. *We show that all distances are preserved under the mapping $w = z\operatorname{cis}\theta$.* To see that this is true, let z_1 and z_2 be arbitrary complex numbers. According to Fact 15.6 part 1, the distance between z_1 and z_2 is equal to $|z_1 - z_2|$.

   Put $c = \operatorname{cis}\theta$. The points z_1 and z_2 are mapped into complex numbers that we denote w_1 and w_2, respectively. In fact, we have $w_1 = cz_1$ and $w_2 = cz_2$. The distance between w_1 and w_2 is $|w_1 - w_2|$ which is equal to

$$|cz_1 - cz_2| = |c(z_1 - z_2)|$$

   which, by Fact 15.6 part 2, is equal to $|c||z_1 - z_2|$. But

$$|c| = |\operatorname{cis}\theta| = 1$$

   because by definition $\operatorname{cis}\theta$ is a complex number of modulus one. Thus, we see

$$|w_1 - w_2| = |z_1 - z_2|$$

   or, in other words, the distance between z_1 and z_2 is the same as the distance between the image points w_1 and w_2.

3. *There are only two types of mappings that have the properties established in 1 and 2,* that is, mappings having the properties that (a) the origin is mapped into itself and (b) all distances are preserved. Such a mapping must be either (a) a *rotation* about the origin or (b) a *reflection* across a line through the origin. We omit the proof of this geometrically plausible fact.

4. *The mapping of the form $w = cz$ cannot be a reflection.* Indeed, suppose it represented a reflection across a certain line l. Then every point on l must be

fixed under the mapping. In particular, there must be a nonzero point z_0 which is mapped into itself. In other words,

$$cz_0 = z_0 \qquad (15.16)$$

Dividing both sides of this equation by the nonzero complex number z_0, we obtain $c = 1$, and, therefore, the mapping must be $w = z$, which is not a reflection.

Since the mapping is not a reflection, then, according to 3, it must be a rotation. $\square$

There is a slight gap in the above proof. We divided by the nonzero complex number z_0 on both sides of equation (15.16). We may know that such a cancellation is correct with real numbers, but how do we know that it is correct with complex numbers? Instead of dealing with this question abstractly, here is an example that is easy to generalize illustrating division of complex numbers.

Example 15.12. Divide $4 + 3i$ by $5 - 2i$.

Solution. We must find the unique complex number z such that

$$(5 - 2i)z = 4 + 3i \qquad (15.17)$$

Multiply both sides of equation (15.17) by $5 + 2i$, obtaining ↞ Dangerous curve!

$$(5 + 2i)(5 - 2i)z = (5 + 2i)(4 + 3i)$$

Carrying out the complex multiplications, we obtain

$$(25 + 4)z \equiv 29z = (20 - 6 + 15i + 8i) \equiv 14 + 23i$$

Now divide both sides of this equation by 29, obtaining

$$z = \frac{14}{29} + \frac{23}{29}i \qquad (15.18)$$

What this shows is that the only possible candidate for the quotient that we are seeking is (15.18). By reversing the steps we have just carried out, we see that (15.18) is, indeed, a solution of equation (15.17).

A similar computation can be used to carry out the division of an arbitrary complex number by an arbitrary nonzero complex number. We can carry out computations with the complex numbers in a manner that is very similar to computations with real numbers. The reason that this is possible is that the complex and real numbers are both examples of a *field*, an abstract structure satisfying a certain list of axioms. One can calculate confidently with real numbers, even without knowing that the real numbers are a field, and one can do the same with complex numbers.

In the above computation, we were more interested in the logical correctness of the steps than in the efficiency of the computation. Let us carry through the same calculation using a method similar to *rationalizing the denominator*. We want to find a form without any i's in the denominator for the following fraction

$$\frac{4 + 3i}{5 - 2i}$$

We proceed as follows:

1. Multiply numerator and denominator by $5 + 2i$, obtaining

$$\frac{4 + 3i}{5 - 2i} \cdot \frac{5 + 2i}{5 + 2i}$$

2. Carry out the multiplications in the numerator and denominator:

$$= \frac{20 + 15i + 8i + 6i^2}{25 + 10i - 10i - 4i^2}$$

3. Collect terms using the fact that $i^2 = -1$:

$$= \frac{20 + 23i - 6}{25 + 4} = \frac{14 + 23i}{29} = \frac{14}{29} + \frac{23}{29}i$$

We can summarize Fact 15.6 and Proposition 15.11 as follows.

Fact 15.13. *Suppose that two complex numbers have the polar representations*

$$r_1 \operatorname{cis} \theta_1 \qquad r_2 \operatorname{cis} \theta_2$$

Then the polar representation of the product is

$$r_1 r_2 \operatorname{cis}(\theta_1 + \theta_2)$$

Note that in this fact we do *not* require the arguments to be nonnegative or less than $360°$.

Summing up, the geometric meaning of complex multiplication is this:

- The *modulus* of the product is the *product* of the moduli.

- The *argument* of the product is the *sum* of the arguments.

Iteration of mappings

Let us consider a recursive mapping process. Suppose that we have a mapping $w = f(z)$ defined in the entire z-plane. Starting with a particular complex number z_0, we define the sequence $z_1, z_2, \ldots$ by the recursion

$$z_{n+1} = f(z_n) \quad n = 1, 2, \ldots \tag{15.19}$$

We will consider instances of this recursion in which the function f depends on a second complex number c. That is, our mapping is of the form $w = f(z, c)$. For example, $f(z, c)$ could be equal to cz; another possibility is $f(z, c) = z^2 + c$. For each complex number c, one of two alternatives must hold.

1. *The points z_n are bounded.* In other words, there exists a number R such that all of the z_n are contained in some circular disk of radius R with its center at the origin; i.e., we have $|z_n| < R$ for all n.

2. *The points z_n are unbounded.* In other words, these points are not all contained in a disk centered at the origin, no matter how large.

For a given mapping $f(z,c)$ and initial value z_0, let $\mathcal{M}$ denote the set of complex numbers c such that alternative 1 holds. In case $f(z,c) = z^2 + c$ and $x_0 = 0$, the set $\mathcal{M}$ is called the *Mandelbrot set* after Benoit Mandelbrot, who was first to notice the remarkable properties of this set, shown in Figure 15.9.

The mapping $w = cz$

Since we have just studied the mapping properties of $w = cz$, let us see if we can discover the set $\mathcal{M}$ for the mapping (15.19) with $f(z,c) = cz$ and the initial condition $z_0 = 1$. There are three cases to consider.

1. $|c| < 1$. The points z_n tend to the origin as n tends to infinity.

2. $|c| = 1$. The points z_n remain on the unit circle $|z| = 1$.

3. $|c| > 1$. The points z_n tend to infinity. More precisely, for any circle, however large, with its center at the origin, all but finitely many of the points z_n are outside the given circle.

The foregoing shows that the set $\mathcal{M}$ for this problem is the unit disk $c \leq 1$. We see that for $f(z) = cz$, the sequence z_n is predictable and straightforward, but we will see that there is a simple choice for the function f that leads to results that are vastly more complex.

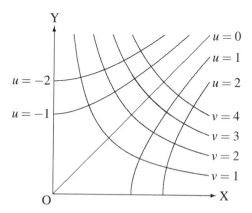

Figure 15.8. The mapping $w = z^2$. Putting $z = x + yi$ and $w = u + vi$, this mapping is equivalent to $u = x^2 - y^2$, $v = 2xy$. The curves in this figure are mapped into straight lines in the w-plane: $u = $ constant, and $v = $ constant. The X and Y axes are mapped into the V-axis in the W-plane.

The mapping $w = z^2 + c$: The Mandelbrot set

The mapping $w = z^2$ is shown in Figure 15.8. In particular, this figure shows the smooth curves in the z-plane that are mapped into horizontal and vertical lines in the w-plane. Geometrically, the mapping stretches the quadrant of the z-plane shown in Figure 15.8 into the entire upper half of the w-plane. Figure 15.8 gives no clue of the strange results just around the corner.

Put $z_0 = 0$, and define a recursion as follows.

$$z_{n+1} = z_n^2 + c \quad n = 1, 2, \ldots \tag{15.20}$$

The set $\mathcal{M}$ for recursion (15.20), called the Mandelbrot set, shown in Figure 15.9, is a very strange set. The boundary of $\mathcal{M}$ is a fractal; that is, the boundary has a similar appearance under any degree of magnification. Under high magnification, the boundary of Figure 15.9 is exceedingly complex with spirals and filaments within spirals and filaments.[5]

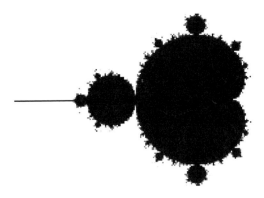

Figure 15.9. The Mandelbrot set.

In summary, the entire number system can be based on the natural numbers. In this process of construction, one builds successively the integers, the rationals, the reals, and finally the complex numbers. In each step, a new class of numbers is defined in terms of the previous class. In this chapter, we have seen two steps in this construction: how the integers are built from the natural numbers and how the complex numbers are built from the reals. Moreover, we have seen the surprising connection of the complex numbers with geometry.

In the next chapter, we introduce the *prime numbers*, a class of integers with many remarkable properties.

Chapter 16

Prime News

Amongst the many changes and alterations which Lycurgus [the traditional law-giver of Sparta] made, the first and of greatest importance was the establishment of the senate .... As for the determinate number of twenty-eight [members] ... perhaps there is some mystery in the number, which consists of seven multiplied by four, and is the first of perfect numbers after six, being, as that is, equal to [the sum of] all its parts.

—PLUTARCH, *Lives*

We define a class of integers called *prime numbers*. The properties of these numbers have engrossed mathematicians for thousands of years. The study of prime numbers is a major part of the branch of mathematics called the *theory of numbers* that deals with the properties of integers.

Prime numbers are defined to be the natural numbers greater than 1 that are divisible only by themselves and 1. A natural number greater than 1 that is not prime is said to be composite. We begin our discussion of prime numbers by stating a few definitions because we need to be very clear about the meanings of several words.

Definition 16.1. If p is a natural number greater than 1 such that p is divisible only by 1 and itself, then we say p is a *prime* number—otherwise we say that p is *composite*.

Note that according to Definition 16.1, 1 is neither prime nor composite.
Figure 16.1 is a list of the first 180 prime numbers, up to 1069.

Definition 16.2. If natural numbers n and m (both greater than 1) have no common divisor greater than 1, then n and m are said to be *relatively prime*.

For example, 24 and 49 are relatively prime because 24 and 49 have no common factor.

There is an important theorem concerning prime numbers that is called the *unique factorization theorem*.

Theorem 16.3 (unique factorization). *Every natural number n greater than 1 can be expressed as a product of prime numbers in exactly one way.*[1]

2	53	127	199	283	383	467	577	661	769	877	983
3	59	131	211	293	389	479	587	673	773	881	991
5	61	137	223	307	397	487	593	677	787	883	997
7	67	139	227	311	401	491	599	683	797	887	1009
11	71	149	229	313	409	499	601	691	809	907	1013
13	73	151	233	317	419	503	607	701	811	911	1019
17	79	157	239	331	421	509	613	709	821	919	1021
19	83	163	241	337	431	521	617	719	823	929	1031
23	89	167	251	347	433	523	619	727	827	937	1033
29	97	173	257	349	439	541	631	733	829	941	1039
31	101	179	263	353	443	547	641	739	839	947	1049
37	103	181	269	359	449	557	643	743	853	953	1051
41	107	191	271	367	457	563	647	751	857	967	1061
43	109	193	277	373	461	569	653	757	859	971	1063
47	113	197	281	379	463	571	659	761	863	977	1069

Figure 16.1. The first 180 prime numbers.

For example, 2, 3, and 5 are primes, and 60 is equal to the product $2 \cdot 2 \cdot 3 \cdot 5$.

The unique factorization theorem gives us a standard way of representing the natural numbers. It is customary to write the prime factors in ascending order and use exponents to indicate repeated factors. Here are a few prime factorizations:

$$30 = 2 \cdot 3 \cdot 5 \quad 36 = 2^2 \cdot 3^2 \quad 64 = 2^6$$

The Infinitude of Prime Numbers

The number of prime numbers less than or equal to n is denoted $\pi(n)$. We can see in Figure 16.2 that, as they increase in size, the prime numbers become more sparse. At some point, do we run out of prime numbers altogether? The answer is that there is no largest prime number—a fact discovered by Euclid around 300 B.C. The proof is short. It is a model of elegance seldom achieved. To understand this proof is to experience an intellectual exhilaration.

Theorem 16.4. *There are infinitely many prime numbers.*

Proof. Let us suppose that there are only finitely many prime numbers, say, there are n of them

$$p_1, p_2, \ldots, p_n \tag{16.1}$$

We deduce a contradiction as follows. Consider the number

$$N = p_1 p_2 p_3 \ldots p_n + 1 \tag{16.2}$$

n	$\pi(n)$	percent
10	4	40.0%
100	25	25.0%
1,000	168	16.8%
10,000	1,229	12.2%
100,000	9,592	9.6%
1,000,000	78,501	7.9%

Figure 16.2. The distribution of prime numbers. The number of prime numbers less than or equal to n is denoted $\pi(n)$. The right column is the percentage of natural numbers less than or equal to n that are prime numbers.

By Theorem 16.3, N must have a prime factorization, and the prime factors must ↩ Dangerous curve! be among the primes listes above in (16.1), because we have supposed that there are no other primes. Suppose that p_r is one of these prime factors so that p_r is a divisor of N. But dividing N by p_r gives a remainder of 1; hence N is not divisible by p_r. Contradiction!   □

This proof is traditionally given in this form, but the idea can be easily converted to a constructive direct proof. That is, we can actually construct an infinite sequence of prime numbers. However, we do not claim that it is possible to construct *all* the prime numbers in this way. In fact, we construct a sequence of numbers that are either prime numbers or composite numbers that are relatively prime to all the earlier members of the sequence. The existence of this sequence implies that there are infinitely many prime numbers. This becomes more clear as we actually construct the sequence.

We proceed as follows. Call this sequence

$$q_1, q_2, q_3, \ldots$$

We use q instead of p because p often stands for *prime* and the numbers in this sequence are not necessarily prime. We start with $q_1 = 2$, which we know is a prime. We define the sequence by a relationship very similar to equation (16.2):

$$q_{n+1} = q_1 q_2 q_3 \cdots q_n + 1 \tag{16.3}$$

An equation like equation (16.3) that defines new members of a sequence in terms of the previous members is called a *recurrence relation*. Let us look at the numbers in this sequence.

$$q_1 = 2$$
$$q_2 = 2 + 1 = 3$$
$$q_3 = 2 \cdot 3 + 1 = 7$$
$$q_4 = 2 \cdot 3 \cdot 7 + 1 = 43$$
$$q_5 = 2 \cdot 3 \cdot 7 \cdot 43 + 1 = 1,807 = 13 \cdot 139$$
$$q_6 = 2 \cdot 3 \cdot 7 \cdot 43 \cdot 1,807 + 1 = 3,263,443$$

$\cdot \quad \cdot \quad \cdot$

Since $2, 3, 7,$ and 43 are primes, one might think that all these numbers are primes. As a matter of fact, $3,263,443$ happens to be a prime, but $1,807$ is not a prime because its prime factorization is $1,807 = 13 \cdot 139$. (We already know that 13 is a prime, and a little checking shows that 139 is also a prime.) The number $1,807$ is not a prime, but there is no difficulty in that. We are certain that q_5 is the *only* member of the

Dangerous curve! ↬ sequence that is divisible by either 13 or 139. In general, the prime factors of each member of the list are all different from the prime factors of all the other members of the list. Thus, each member of the list contributes at least one new prime number, and this implies that there are infinitely many prime numbers.

The following fact is an extension of Euclid's theorem.

Fact 16.5. *Let a and b be relatively prime natural numbers. Then there exist infinitely many natural numbers n such that $an + b$ is a prime number.*

For example, putting $a = 4$ and $b = 1$, there are infinitely many primes in the sequence

$$1, 5, 9, 13, 17, \ldots$$

The Prime Number Theorem

From Figure 16.2, we know that as we examine larger and larger natural numbers, the primes become more and more sparse. This fact intrigued mathematician Karl Friedrich Gauss (1777–1855) as a young schoolboy. By counting lists of prime num-

See the discussion of logarithms on page 129. The natural logarithm of n is written $\ln n$ and is equal to the base-ten logarithm $\log n$ multiplied by $2.302585\ldots$.

bers, Gauss arrived at a conjecture as to how the prime numbers are distributed. (Recall that $\pi(n)$ is the number of primes less than or equal to n.) Gauss conjectured that $\pi(n)$ is approximately equal to $n/\ln n$. More precisely, Gauss conjectured that

$$\frac{\pi(n)}{n/\ln n} \quad \text{tends to 1 as } n \text{ becomes large}$$

In other words, Gauss conjectured that the numbers in the right column of Figure 16.3 get closer and closer to 1 as n gets larger and larger. Gauss's conjecture is now called the *prime number theorem*. Although Figure 16.3 provides only meager support, it turns out that the prime number theorem is true. Gauss never published his conjecture, presumably because he was unable to prove it. This is not the only time

n	$\pi(n)$	$n/\ln n$	$\frac{\pi(n)}{n/\ln n}$
10	4	4.343	0.921
100	25	21.71	1.151
1,000	168	144.76	1.161
10,000	1,229	1,085.7	1.132
100,000	9,592	8,685.9	1.104
1,000,000	78,501	72,3824	1.085

Figure 16.3. The prime number theorem asserts that the numbers in the right column tend to 1 as n grows large.

that Gauss failed to publish important work because he felt that it was insufficiently polished.

Later in 1808, the French mathematician Adrien Marie Legendre (1752–1833) published a similar conjecture—that $\pi(n)$ for large n is approximately equal to

$$\frac{n}{\ln n + B}$$

where B is equal to $-1.08366\dots$. It is now known that Legendre could have improved his conjecture by giving B the value -1.

Forty-one years after the death of Gauss, the prime number theorem was finally proved independently and simultaneously in 1896 by the French mathematician Jacques Hadamard (1865–1963) and the Belgian Charles-Jean de la Vallée Poussin (1866–1962). Their proofs are extremely difficult and involve the theory of functions of a complex variable, a topic with no apparent connection to the prime numbers. For more than 50 years, mathematicians sought an "elementary" proof of the prime number theorem, a proof not involving functions of a complex variable. Such a proof would be more in accord with Ockham's razor.[2] In 1949, the Norwegian mathematician Atle Selberg (1917–) and the Hungarian mathematician Paul Erdős (1913–96) discovered an elementary proof of the prime number theorem. In 1950, Selberg received the prestigious Fields Medal for this work. As a measure of difficulty, consider that in 1950, the author of this book attended a course of about 30 lectures devoted to an exposition of the new "elementary" proof of the prime number theorem.[3]

Unsolved problems involving prime numbers

There is an unsolved problem that appears to be related to Theorem 16.4. This problem concerns *prime twins*, pairs of prime numbers, like 5, 7 and 11, 13, that differ by two. It is unknown whether there are infinitely many such prime twins.

The prime twins conjecture is one of many problems of number theory that are easy to state but difficult to solve. Indeed, many such problems remain unsolved despite intense effort. One such unsolved problem, the *Goldbach conjecture*, asserts that every even natural number greater than 2 is the sum of two prime numbers. This problem was first stated in about 1742, in a letter from Christian Goldbach (1690–1764) to Leonhard Euler. Goldbach was a German mathematics teacher; Czar Peter II, grandson of Peter the Great, was one of his pupils. In the 1951 movie, *No Highway in the Sky* with James Stewart and Marlene Dietrich, Stewart was an eccentric aeronautical scientist who worked on the Goldbach conjecture. That movie contains an accurate statement of the conjecture. The Goldbach conjecture remains an unsolved problem despite the best efforts of generations of mathematicians. The difficulty of the problem stems from the fact that prime numbers, by definition, relate to the representation of natural numbers as *products* whereas the Goldbach conjecture seeks to represent natural numbers as *sums* of prime numbers.

Formulas that generate prime numbers

There are simple formulas that produce many primes. Here are two examples.

$$f(n) = n^2 - n + 41$$
$$g(n) = n^2 - 79n + 1601$$

By coincidence, 1601 is the year that mathematician Pierre de Fermat was born. For $n = 1, 2, 3, \ldots, 40$, $f(n)$ is a prime, and for $n = 1, 2, 3, \ldots, 79$, $g(n)$ is a prime. However, $f(41)$ and $g(80)$ are composite; in fact, both are equal to 41^2.

The strangest formula that generates prime numbers must be the following. There exists a constant $\theta = 1.3064\ldots$, known as Mill's constant, such that the integer part of

$$\theta^{3^n}$$

is a prime number for *every* natural number n.

Fermat and Mersenne primes

Suppose that one wants to find primes that are either one more or one less than a power of 2—primes of the form $2^n \pm 1$ where n is a natural number. The following facts can be proved with elementary algebra.

Fact 16.6. *If for some natural number n, $2^n + 1$ is a prime, then n must be a power of 2, that is, n must be one of the natural numbers $1, 2, 4, 16, \ldots$. In other words, the prime number must be of the form $2^{2^m} + 1$ for some nonnegative integer m.*

Fact 16.7. *If for some natural number n, $2^n - 1$ is a prime, then n must be a prime.*

Fermat numbers

Numbers of the type $2^{2^m} + 1$ were first studied by Pierre de Fermat,[4] and hence are called Fermat numbers. From Fact 16.6, it follows that all primes of the form $2^n + 1$

must be Fermat numbers. Let F_n denote the nth Fermat number. The first few Fermat numbers are

$$F_0 = 2^{2^0} + 1 = 2^1 + 1 = 3$$
$$F_1 = 2^{2^1} + 1 = 2^2 + 1 = 5$$
$$F_2 = 2^{2^2} + 1 = 2^4 + 1 = 17$$
$$F_3 = 2^{2^3} + 1 = 2^8 + 1 = 257$$
$$F_4 = 2^{2^4} + 1 = 2^{16} + 1 = 65,537$$
$$F_5 = 2^{2^5} + 1 = 2^{32} + 1 = 4,294,967,297$$

Fermat noticed that the numbers 3, 5, 17, 257, and 65,537 are all prime numbers, and he conjectured that the Fermat numbers are *all* prime. This turned out to be incorrect; in fact, Leonhard Euler discovered that F_5 is divisible by 641. Currently, F_0 through F_4 are the only Fermat numbers that are known to be primes.

Fermat numbers are more than just a historical curiosity. There is a very strange connection with a problem that concerned the ancient Greek geometers. That problem is the construction of a regular polygon using a ruler and compass only. A regular n-sided polygon, also called an n-gon, has the property that all its sides are of equal length and all of its angles are equal. The ancient Greek geometers knew how to construct, using a ruler and compass only, regular polygons having three or five sides, but they did not know how to make this construction for any regular polygons with a prime number of sides larger than five. The problem had to wait for thousands of years. At the age of seventeen, Gauss showed that the construction of a regular polygon with a prime number p of sides is possible if and only if p is a Fermat number. The third Fermat number, seventeen, had a special significance for Gauss. At the age of *seventeen*, he discovered that a regular polygon with *seventeen* sides can be constructed using only ruler and compass. At Göttingen, the German university at which Gauss was professor and unquestioned leader of the mathematical world for most of his long life, there is a statue of Gauss on a pedestal with *seventeen* sides.

The study of prime numbers may at first seem to be the purest of pure mathematics—out of reach of any technological application. That might have been true 50 years ago, but today the study of prime numbers plays a role in digital signal processing and in cryptography.

Now let us take a closer look at the primes of the type mentioned in Fact 16.7.

Mersenne numbers

Numbers of the form $M_p = 2^p - 1$ where p is a prime number are called Mersenne numbers after the French mathematician Marin Mersenne (1588–1648). Fact 16.7 asserts that a number of the form $2^p - 1$ cannot be prime unless p is a prime; however, not all prime values of p generate Mersenne primes. Special methods exist for determining the primality of Mersenne numbers, and the largest known prime numbers are Mersenne primes. Large prime numbers play a role in the most advanced encryption schemes, which are called *public-key ciphers*. (See Chapter 18.)

The largest known prime number can expect to enjoy that status for only a short while because every few months a new largest prime is discovered—thanks to the Great Internet Mersenne Prime Search (GIMPS), an organization of thousands of number theory enthusiasts who use their computers cooperatively for this task. On January 27, 1998, the 909,526 digit number $2^{3,021,377} - 1$ became the largest known prime number and the thirty-seventh known Mersenne prime. The first few Mersenne primes are shown in Figure 16.4.

$$
\begin{array}{lll}
M_2 = \ 3 & M_7 = \quad\ \ 127 & M_{19} = \qquad\ 524,287 \\
M_3 = \ 7 & M_{13} = \quad 8,191 & M_{31} = 2,147,483,647 \\
M_5 = 31 & M_{17} = 131,071 &
\end{array}
$$

Figure 16.4. Mersenne primes.

Perfect numbers

The Mersenne primes have a relationship to the *perfect numbers*, a title bestowed by the Pythagoreans. Numerology often motivated their scientific investigations, and, in particular, they believed that these numbers have magical properties. The perfect numbers are defined as follows.

Definition 16.8. A number is said to be *perfect* if it is equal to the sum of all its proper divisors.

In this definition, the *proper divisors* include 1 and exclude the number itself. For example, the proper divisors of 24 are

$$1, 2, 4, 8, 3, 6, \text{and } 12$$

and since the sum of these numbers is 36, 24 is not a perfect number. It is easy to check that 6 and 28 are perfect numbers.

The connection between Mersenne primes and perfect numbers, known to Euclid, is contained in the following proposition.

Proposition 16.9. *Let p be a natural number. Put P equal to*

$$P = 2^{p-1}(2^p - 1) \tag{16.4}$$

and put q equal to

$$q = 2^p - 1 \tag{16.5}$$

Then P is a perfect number if and only if q is a prime.

Proof. We need to use Theorem 3.8, the formula for the sum of a geometric progression. More specifically, we use Example 3.10, which follows Theorem 3.8. Suppose that equation (16.4) is true and (16.5) is a prime. We can enumerate the divisors of *P*. First there are the powers of 2:

$$1 = 2^0, 2 = 2^1, 2^2, 2^3, \ldots, 2^{p-1}$$

We apply Example 3.10 to find that the sum of these divisors is $2^p - 1$. The other ↶ Dangerous curve! divisors of *P* are the following:

$$q, 2q, 2^2q, 2^3q, \ldots, 2^{p-2}q$$

We apply Example 3.10 again to find that the sum of these divisors is equal to $q(2^{p-1} - 1)$. Since *q* is a prime, there are no other divisors of *P*. Now we see that the sum of all of the divisors of *P* is

$$2^p - 1 + q(2^{p-1} - 1)$$

But now use equation (16.5), the definition of *q*. The previous formula becomes

$$q + q2^{p-1} - q$$

which from equations (16.4) and (16.5) is equal to *P*. This proves the *if* part of Proposition 16.9. To prove the *only if* part, we observe that if *q* is not a prime, there are other divisors than the ones enumerated above. Consequently, the proper divisors would have a sum greater than *P*.   □

Since M_2, M_3, M_5, and M_7 are primes, the first four perfect numbers are as follows:

$$P_2 = 2 \cdot (2^2 - 1) = 6$$
$$P_3 = 2^2 \cdot (2^3 - 1) = 28$$
$$P_5 = 2^4 \cdot (2^5 - 1) = 496$$
$$P_7 = 2^6 \cdot (2^7 - 1) = 8,128$$

Q. Are all of the perfect numbers generated in this way?

A. All of the even *perfect numbers are generated in this way. It is not known if there exist odd perfect numbers.*

Multiperfect numbers

A natural number is *multiperfect* if the sum of its divisors is a multiple of itself. On May 13, 1997, Ron Sorli announced the discovery of the first 10-perfect number, a number such that the sum of its proper divisors is equal to the number multiplied by nine.[5] This number is approximately equal to 2.868798×10^{923}, a number with 924 decimal digits.

We have discussed the infinitude of prime numbers and certain special classes of primes. In the next chapter, we will see another topic of the theory of numbers. We begin with a discussion of a problem from an ancient Indian mathematician, a puzzle concerning a group of travelers who come upon piles of fruit in the forest.

Chapter 17

The Unknown Division

Ah, but my Computations, people say,
Reduced the year to better reckoning?—Nay,
'Twas only striking from the Calendar
Unborn To-morrow, and dead Yesterday.

—OMAR KHAYYÁM, *Rubáiyát*

Our discussion of the Pythagorean triples on page 180 introduced diophantine equations—equations that require solutions in integers (sometimes rational numbers).
Some of the most difficult problems in mathematics—for example, Fermat's last
theorem (Theorem 12.10)—are diophantine equations. We begin this chapter with
an ancient Indian puzzle that leads to a type of diophantine equation for which there
exist elegant methods of solution. We will see a solution by means of the Euclidean
algorithm, a procedure of fundamental importance in number theory. We interpret
this solution using continued fractions—an alternative basis for the arithmetic of real
numbers—and briefly explore this strange mathematical landscape. We will see that
continued fractions provide a variant representation of numbers. Finally, continued
fractions give insight into the meaning of leap year.

A Linear Diophantine Problem

The following puzzle is from a ninth-century Indian mathematician, the Jain monk
Mahaviracarya.[1]

Problem 17.1. In the forest, 37 numerically equal heaps of wood apples were seen
by travelers. After 17 fruits were removed, the remainder was divided evenly among
79 persons. What is the share obtained by each?

We can translate this problem into an algebraic equation as follows. There are
two unknown quantities, the number of apples in each heap and the share of apples
for each person; let us denote them x and y, respectively. We can then express the
total number of apples in two different ways.

1. There are 37 heaps, each with x apples, a total of $37x$.

2. Each of the 79 travelers receives y apples. Adding the 17 discarded fruits, the original total must have been $79y + 17$.

By equating the total apples in 1 and 2, we obtain

$$37x = 79y + 17$$

We get a more standard form of the equation by putting the variables on one side and the constant on the other by subtracting $79y$ from both sides.

$$37x - 79y = 17 \qquad (17.1)$$

This equation is a mathematical model for Problem 17.1 in the same sense that equation (2.2) was a mathematical model for a problem involving the hands of a clock. Solving Problem 17.1 is equivalent to finding nonnegative integers x, y that solve equation (17.1).

Part of the charm of Problem 17.1 is that there appears to be insufficient information to determine a solution. Indeed, this problem has more than one solution—in fact, infinitely many. A problem of this sort is called *indeterminate* because a single equation involving two variables does not ordinarily lead to a unique solution.

> Any equation of the form $ax + by = c$ where a, b, and c are constants is represented graphically by a straight line. This correspondence between algebraic equations and geometric figures introduces a powerful method called *analytic geometry*, also called Cartesian geometry in honor of its discoverer, the French philosopher and mathematician René Descartes (1596–1650).

The graph of points (x, y) that satisfy equation (17.1) is a straight line, specifically the line shown in Figure 17.1. Because the solutions to equation (17.1) lie on a straight line, we are concerned here with a *linear* diophantine problem. We will not give a proof that the graph of equation (17.1) is a straight line.[2]

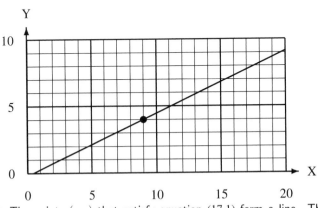

Figure 17.1. The points (x, y) that satisfy equation (17.1) form a line. The points—called lattice points—for which both x and y are integers are the intersections of the horizontal and vertical lines. In the portion of the X-Y plane shown, the lattice point at $x = 9, y = 4$ appears to lie on the line; and indeed, a computation verifies that these values of x and y satisfy equation (17.1).

Graphical solution

The points in Figure 17.1 where both x and y are integers are called lattice points. Geometrically, Problem 17.1 requires us to find lattice points on a certain line. It appears that the point $x = 9, y = 4$ is on the line, and we confirm this fact by verifying that these values satisfy equation (17.1). This fact could never be demonstrated simply by improving the accuracy of our diagram. The graphical method is limited to discovering likely points that we then check by numerical calculation. This method helps us find the point $(9, 4)$, but it cannot help us to find *all* solutions.

Algebraic solution

The algebraic solution of Problem 17.1 is rather lengthy, but we obtain a complete   ↤ Dangerous curve! solution in this way. This method is like taking apart and putting back together a nested set of Russian babushka dolls of graduated size, each fitting within the one immediately larger. Let us assume that we have a solution of Problem 17.1 and that x and y are nonnegative integers that solve equation (17.1). We begin by solving equation (17.1) for x as follows.

$$x = \frac{79y + 17}{37} \tag{17.2}$$

Now using the fact that

$$\frac{79}{37} = 2 + \frac{5}{37}$$

equation (17.2) becomes

$$x = 2y + \frac{5y + 17}{37} \tag{17.3}$$

Now we see that, since the last term on the right side of equation (17.3) is an integer, there must be an integer n such that

$$5y + 17 = 37n \tag{17.4}$$

Solving for y, we find

$$y = \frac{37n - 17}{5} = 7n + \frac{2n - 17}{5} \tag{17.5}$$

Now since $2n - 17$ must be divisible by 5, there must be an integer m such that

$$2n - 17 = 5m \tag{17.6}$$

Solving for n, we find

$$n = \frac{5m + 17}{2}$$

Again, since $5m + 17$ must be divisible by 2, there must be an integer k such that

$$5m + 17 = 2k \tag{17.7}$$

Solving for k, we obtain

$$k = \frac{5m + 17}{2} = 2m + \frac{m + 17}{2}$$

Now we see that the right side is an integer if and only if m is an odd integer. In other words, there exists an integer j such that

$$m = 2j + 1 \tag{17.8}$$

Now substitute equation (17.8) back into equation (17.7), obtaining

$$5(2j + 1) + 17 = 10j + 5 + 17 = 10j + 22 = 2k \tag{17.9}$$

Now use equation (17.9) to eliminate k from equation (17.7), obtaining

$$5m + 17 = 10j + 22$$
$$5m = 10j + 5 \tag{17.10}$$

Now use equation (17.10) to eliminate m from equation (17.6), obtaining

$$2n - 17 = 10j + 5$$
$$2n = 10j + 22$$
$$n = 5j + 11 \tag{17.11}$$

Now use equation (17.11) to eliminate n from equation (17.5)

$$y = 7n + \frac{2n - 17}{5} \tag{17.12}$$
$$= 7(5j + 11) + 2j + 1 = 37j + 78$$

Now use equations (17.12), (17.4), and (17.11) to eliminate y from equation (17.3), obtaining

$$x = 2y + \frac{5y + 17}{37}$$
$$= 2(37j + 78) + n = 74j + 156 + 5j + 11$$
$$= 79j + 167$$

Let us state the result we have obtained from this rather long computation as follows.

Proposition 17.2. *For every integer j, the integers*

$$x = 79j + 167 \qquad y = 37j + 78 \tag{17.13}$$

solve equation (17.1). Moreover, all integral solutions of equation (17.1) are obtained in this way.

The solution $x = 9$, $y = 4$, found in Figure 17.1 is obtained by putting $j = -2$, but now we see that this solution is only one among infinitely many. Only the solutions with nonnegative x and y are meaningful for the original problem concerning the travelers and the heaps of fruit. These solutions are obtained by putting $j \geq -2$. The next larger solution is obtained by putting $j = -1$ in equation (17.13), which yields $x = 88$ and $y = 41$—each heap has 88 apples, and each traveler gets 41 of them.

In the next section, we introduce the Euclidean algorithm, a concept—interesting in its own right—that will enable us to solve such diophantine problems in a more systematic way.

The Euclidean Algorithm

The division algorithm (Theorem 8.2) asserts that after dividing an integer by a natural number we obtain a quotient and a remainder. Please remember that the division algorithm refers to *division of whole numbers ignoring any fractional part of the quotient*, a more primitive kind of division than the familiar one.

Recall the definition of divisibility (Definition 4.5). In terms of the division algorithm, if division of a by b yields a remainder $r = 0$, then b divides a. There is a simple but useful property of divisibility that we will now state as a proposition.

Fact 17.3. *Let a, b, and c be natural numbers. If both a and b are divisible by c, then $a + b$ is divisible by c.*

We omit the proof. We continue with two more definitions.

Definition 17.4. Let n and m be natural numbers. The largest natural number that divides both n and m is called the *greatest common divisor* (g.c.d.) of n and m and is denoted (n, m).

For example, we have $(24, 44) = 4$ because 4 is the largest natural number that divides both 24 and 44. Note that, according to Definition 16.2, natural numbers n and m (both greater than 1) are relatively prime if and only if $(n, m) = 1$.

The Euclidean algorithm is a procedure for finding the g.c.d. of two numbers. Let us suppose that we want to find the g.c.d. of natural numbers a and b. Then the division algorithm Theorem 8.2 asserts that there exist nonnegative integers q_1 and r_1 such that

$$a = q_1 b + r_1 \qquad 0 \leq r_1 < b \qquad\qquad (17.14)$$

If $r_1 > 0$, then we continue by repeatedly applying the division algorithm until we find a remainder that is equal to zero.

$$b = q_2 r_1 + r_2 \qquad 0 \leq r_2 < r_1 \qquad\qquad (17.15)$$
$$r_1 = q_3 r_2 + r_3 \qquad 0 \leq r_3 < r_2 \qquad\qquad (17.16)$$
$$r_2 = q_4 r_3 + r_4 \qquad 0 \leq r_4 < r_3 \qquad\qquad (17.17)$$

$\cdot\quad\cdot\quad\cdot$

Notice that if a remainder is positive, then the next remainder is smaller. Eventually, we will reach a remainder that is zero; let us say $r_{n+1} = 0$. At this point the process comes to an end.

$$\cdot \quad \cdot \quad \cdot$$

$$r_{n-3} = q_{n-1}r_{n-2} + r_{n-1} \qquad 0 \le r_{n-1} < r_{n-2} \qquad (17.18)$$

$$r_{n-2} = q_n r_{n-1} + r_n \qquad 0 \le r_n < r_{n-1} \qquad (17.19)$$

$$r_{n-1} = q_{n+1}r_n \qquad\qquad\qquad\qquad\qquad (17.20)$$

The entire procedure of successive applications of the division algorithm is called the Euclidean algorithm. An important fact concerning the Euclidean algorithm is stated in the following proposition.

Theorem 17.5. *The remainder r_n is the g.c.d. of a and b. Moreover, any common divisor of a and b must also be a divisor of r_n.*

This theorem is needed to give a proof—we will not do so—of the unique factorization theorem, Theorem 16.3. The method of the proof of Theorem 17.5 is the repeated application of Fact 17.3; (1) starting with equation (17.20) and proceeding backward to equation (17.14), and (2) starting with equation (17.14) and proceeding forward to equation (17.20).

Proof. From equation (17.20) it follows that r_{n-1} is divisible by r_n. From equation (17.19) and Fact 17.3 it follows that r_{n-2} is divisible by r_n. From equation (17.18) and Fact 17.3 it follows that r_{n-3} is divisible by r_n, and so on. Finally, we find that both a and b are divisible by r_n. This shows that r_n is a common divisor of a and b.

Dangerous curve! ↪   We still must show that r_n is the *greatest* common divisor of a and b. It is sufficient to show that every common divisor of a and b is also a divisor of r_n. Suppose that d is a divisor of both a and b. Then from equation (17.14) and Fact 17.3 it follows that r_1 is divisible by d; from equation (17.15) that r_2 is divisible by d; and from equation (17.16) that r_3 is divisible by d. Eventually, we see that r_{n-1} is divisible by d. Then from equation (17.20) it follows that, as claimed, r_n is divisible by d, and we have shown that r_n is the g.c.d. of a and b. Furthermore, we have shown that if d is an arbitrary common divisor of a and b, then d must also be a divisor of r_n.   ☐

Example 17.6. Compute the g.c.d. of 1633 and 713.

Solution. The Euclidean algorithm proceeds as follows. In numerical computations of the Euclidean algorithm we will write the remainders and the initial numbers a and b in boldface and the quotients in normal Roman type.

$$\mathbf{1633} = 2 \cdot \mathbf{713} + \mathbf{207}$$

$$\mathbf{713} = 3 \cdot \mathbf{207} + \mathbf{92}$$

$$\mathbf{207} = 2 \cdot \mathbf{92} + \mathbf{23}$$

$$\mathbf{92} = 4 \cdot \mathbf{23}$$

We conclude that 23 is the g.c.d. of 1633 and 713—in other words,

$$(1633, 713) = 23$$

It is easy to verify without using the Euclidean algorithm that 23 is a divisor of both of these numbers because $1633 = 23 \cdot 71$ and $713 = 23 \cdot 31$. One can verify that there are no larger common divisors.

Example 17.7. Compute the g.c.d. of 79 and 37.

Solution. The Euclidean algorithm proceeds as follows.

$$\mathbf{79} = 2 \cdot \mathbf{37} + \mathbf{5}$$
$$\mathbf{37} = 7 \cdot \mathbf{5} + \mathbf{2}$$
$$\mathbf{5} = 2 \cdot \mathbf{2} + \mathbf{1}$$
$$\mathbf{2} = 2 \cdot \mathbf{1}$$

We conclude that 79 and 37 are relatively prime—in other words, $(79, 37) = 1$.

The numbers 79 and 37 in the preceding example also appear in Problem 17.1. In fact, we will see that Example 17.7 provides assistance in the solution of Problem 17.1, but first we consider a closely related problem.

Problem 17.8. Find integers X and Y such that

$$37X - 79Y = 1 \tag{17.21}$$

We have shown that 37 and 79 are relatively prime. If this were not the case—if the two numbers had a common factor—then Problem 17.8 would have no solution.

If we find X and Y that solve Problem 17.8, then we can also find an integral solution of equation (17.1). In fact,

$$x = 17X \quad y = 17Y \tag{17.22}$$

is a solution of equation (17.1)—but not necessarily a nonnegative solution.

The computation in Example 17.7 enables us to find a solution of Problem 17.8. First we modify the calculation of Example 17.7 by solving for each of the remainders.

$$\mathbf{5} = \mathbf{79} - 2 \cdot \mathbf{37} \tag{17.23}$$
$$\mathbf{2} = \mathbf{37} - 7 \cdot \mathbf{5} \tag{17.24}$$
$$\mathbf{1} = \mathbf{5} - 2 \cdot \mathbf{2} \tag{17.25}$$

The boldface convention helps to keep track of the remainders in the following calculation. It is helpful to pretend that the boldface numbers are so illegible that we are not able to carry out any addition or multiplication, indeed any arithmetic, that involves the boldface numbers. We use equation (17.24) to eliminate the remainder **2** from equation (17.25), obtaining

$$1 = 5 - 2 \cdot (37 - 7 \cdot 5)$$
$$= 5 - 2 \cdot 37 + 14 \cdot 5 \qquad (17.26)$$
$$= -2 \cdot 37 + 15 \cdot 5$$

Dangerous curve! ↪

Now use equation (17.23) to eliminate the remainder **5** from equation (17.26), obtaining

Recall the order of precedence of arithmetic operations. (See page 30.)
Recall the algebraic technique for *expanding* a parenthetical expression.

$$1 = -2 \cdot 37 + 15 \cdot 5$$
$$= -2 \cdot 37 + 15 \cdot (79 - 2 \cdot 37)$$
$$= 15 \cdot 79 - 2 \cdot 37 - 30 \cdot 37 \qquad (17.27)$$
$$= -32 \cdot 37 + 15 \cdot 79$$

And now we are finished because

$$1 = -32 \cdot 37 + 15 \cdot 79$$

is equation (17.21) with $X = -32$ and $Y = -15$. Substituting these values into equations (17.22), we find the following solution of equation (17.1):

$$x = 17X = 17 \cdot (-32) = -544$$
$$y = 17Y = 17 \cdot (-15) = -255 \qquad (17.28)$$

Since x and y are negative, solution (17.28) of equation (17.1) is not meaningful for Problem 17.1. Nevertheless, this solution can be obtained from the previous solution (17.13) by putting $j = -9$. Furthermore, from this one solution (17.28) of equation (17.1) we can find *all* solutions by the following device.

Proposition 17.9. *Let a, b, and c be integers such that a and b are relatively prime, that is, such that $(a, b) = 1$. Let x_0, y_0 be an integral solution of*

$$ax + by = c \qquad (17.29)$$

Then for every integer n,

$$x = x_0 - bn, \ y = y_0 + an \qquad (17.30)$$

is also an integral solution of equation (17.29); moreover, all integral solutions are obtained in this way.

Proof. Substitution of (17.30) into equation (17.29) gives

$$a(x_0 - bn) + b(y_0 + an) = ax_0 + by_0 - abn + abn$$
$$= ax_0 + by_0 = c$$

The last equality follows because x_0, y_0 is assumed to be a solution of equation (17.29). This shows that (17.30) is an integral solution of equation (17.29) for every integer n. Now we must show that all solutions are generated in this way.

Let x_1, y_1 be an arbitrary integral solution of equation (17.1). Then

$$ax_0 + by_0 = ax_1 + by_1$$

This relationship can also be written

$$a(x_0 - x_1) = b(y_1 - y_0) \qquad (17.31)$$

Now, since the left side is divisible by a, the right side must also be divisible by a. Since a and b are relatively prime, 1 is the only divisor of a that also divides b. It follows that $y_1 - y_0$ must be divisible by a. In other words, there exists an integer n such that $y_1 - y_0 = an$. Adding y_0 to both sides of this equation, we obtain

$$y_1 = y_0 + an$$

Substituting $y_1 - y_0 = an$ in equation (17.31), we obtain $a(x_0 - x_1) = abn$. Dividing both sides of this equation by a (which must be nonzero since we assume a and b are relatively prime) and solving for x_1, we obtain

$$x_1 = x_0 - bn$$

and we are done.   □

In equation (17.1) we have $a = 37$ and $b = -79$. From Proposition 17.9 it follows that every solution of equation (17.1) is of the form

$$x = -544 + 79n \quad y = -255 - 37n$$

for suitably chosen integer n.

Continued Fractions

In this section we will see a different way of carrying out the Euclidean algorithm that leads to the strange world of *continued fractions*. We begin by rewriting the calculation of Example 17.7.

$$\frac{79}{37} = 2 + \frac{5}{37} \qquad \frac{37}{5} = 7 + \frac{2}{5} \qquad \frac{5}{2} = 2 + \frac{1}{2}$$

Finally, we use the above to obtain the continued fraction on the right side of this chain of equalities:

$$\frac{79}{37} = 2 + \frac{5}{37} = 2 + \frac{1}{\frac{37}{5}} = 2 + \frac{1}{7 + \frac{2}{5}} = 2 + \frac{1}{7 + \frac{1}{\frac{5}{2}}} = 2 + \frac{1}{7 + \frac{1}{2 + \frac{1}{2}}} \qquad (17.32)$$

In general, an expression of the form

$$a_0 + \cfrac{1}{a_1 + \cfrac{1}{a_2 + \cfrac{1}{a_3 + \cdots}}} \qquad (17.33)$$

in which the a's are integers is called a *continued fraction*. All the a's except a_0 must be positive. We will only consider continued fractions like (17.33) in which all the numerators are 1's. This special type of continued fraction is sometimes called a *simple* continued fraction. In the remainder of this chapter, we discuss only simple continued fractions; for brevity we generally omit the term *simple*. Since formula (17.33) is typographically awkward, we use the notation

$$[a_0; a_1, a_2, a_3, \ldots] \qquad (17.34)$$

to represent the above continued fraction. Note that the integer part of the continued fraction, a_0, is separated by a semicolon. The a's must be integers, and all of them except a_0 must be nonzero.

The anatomy of a simple continued fraction

In formula (17.34), the numbers $a_0, a_1, a_2, \ldots$ are called *partial quotients*. They are the quotients that appear in the Euclidean algorithm. The truncated continued fractions formed from initial subsequences of the partial quotients of (17.34)

$$[a_0;], \ [a_0; a_1], \ [a_0; a_1, a_2], \ [a_0; a_1, a_2, a_3], \ldots$$

are called the *convergents* of the continued fraction. The convergents are rational numbers that we represent as (improper) fractions:

$$[a_0;] = \frac{p_0}{q_0}, \quad [a_0; a_1] = \frac{p_1}{q_1}, \quad [a_0; a_1, a_2] = \frac{p_2}{q_2}, \quad [a_0; a_1, a_2, a_3] = \frac{p_3}{q_3}, \quad \ldots$$

By definition, we put $q_0 = 1$.

Example 17.10. Compute the convergents of the continued fraction (17.32), which represents $79/37$.

Solution. We will find an easier way to make this computation using equation (17.35), but to clarify the meaning of this computation, this once we will avoid using that shortcut. Using the notation of (17.34), we have

$$\frac{79}{37} = [2; 7, 2, 2]$$

The convergents are as follows:

$$\frac{p_0}{q_0} = [2;] = \frac{2}{1} \qquad\qquad = 2.0000$$

$$\frac{p_1}{q_1} = [2;7] = 2 + \frac{1}{7} = \frac{15}{7} \qquad\qquad \approx 2.1429$$

$$\frac{p_2}{q_2} = [2;7,2] = 2 + \frac{1}{7 + \frac{1}{2}} = 2 + \frac{1}{\frac{15}{2}} = 2 + \frac{2}{15} = \frac{32}{15} \qquad \approx 2.1333$$

$$\frac{p_3}{q_3} = [2;7,2,2] = 2 + \cfrac{1}{7 + \cfrac{1}{2 + \frac{1}{2}}} = 2 + \cfrac{1}{7 + \cfrac{1}{\frac{5}{2}}} = 2 + \cfrac{1}{7 + \frac{2}{5}}$$

$$= 2 + \frac{1}{\frac{37}{5}} = 2 + \frac{5}{37} = \frac{79}{37} \qquad\qquad \approx 2.1351$$

We can observe in this example several phenomena that are true generally for the sequence of convergents of any simple continued fraction. We will not give proofs of these facts.

1. The successive convergents approach numerically the value of the last convergent $79/37$. (Recall that the last convergent is the entire continued fraction with no truncation.)

2. The convergents are fractions, proper or improper, in lowest terms. That is, the numerator and denominator are always relatively prime.

3. The convergents are alternately smaller and larger than the fraction that generates them. The even convergents are smaller and the odd convergents are larger.

4. The penultimate convergent $32/15$—the next to last—contains numbers that we have seen before. In fact, we found on page 244 that $X = -32, Y = -15$ is a solution of equation (17.21). This is not an accident.

Fact 17.11. *Let a and b be natural numbers that are relatively prime. Let p/q be the penultimate convergent for a/b. Then $x = q$, $y = p$ satisfy the equation $ax - by = \pm 1$, where the sign is plus or minus depending on whether the total number of partial quotients, including possibly an initial partial quotient zero, is even or odd.*

There is a much easier way to compute the convergents of a continued fraction than the one that we used above in Example 17.10.

Fact 17.12. *The convergents of a continued fraction satisfy the following recurrence relation.*

$$\frac{p_{n+1}}{q_{n+1}} = \frac{a_{n+1}p_n + p_{n-1}}{a_{n+1}q_n + q_{n-1}} \qquad (17.35)$$

Example 17.13. Use Fact 17.12 to compute the convergents of $79/37$.

Solution. According to (17.32), $79/37 = [2;7,2,2]$. Therefore, we must apply equation (17.35) with

$$a_0 = 2 \quad a_1 = 7 \quad a_2 = 2 \quad a_3 = 2$$

To start this recurrence, we need to compute the first two convergents as before, obtaining

$$\frac{p_0}{q_0} = \frac{2}{1} \qquad \frac{p_1}{q_1} = \frac{15}{7}$$

Now we apply equation (17.35), obtaining

$$n = 1: \qquad \frac{p_2}{q_2} = \frac{a_2 p_1 + p_0}{a_2 q_1 + q_0} = \frac{2 \cdot 15 + 2}{2 \cdot 7 + 1} = \frac{32}{15}$$

$$n = 2: \qquad \frac{p_3}{q_3} = \frac{a_3 p_2 + p_1}{a_3 q_2 + q_1} = \frac{2 \cdot 32 + 15}{2 \cdot 15 + 7} = \frac{79}{37}$$

Example 17.14. Compute the convergents of $[2;3,4,5,6]$.

Solution. The first two convergents are:

$$\frac{p_0}{q_0} = \frac{2}{1} \qquad \frac{p_1}{q_1} = 2 + \frac{1}{3} = \frac{7}{3}$$

Now apply equation (17.35):

$$n = 1: \qquad \frac{p_2}{q_2} = \frac{a_2 p_1 + p_0}{a_2 q_1 + q_0} = \frac{4 \cdot 7 + 2}{4 \cdot 3 + 1} = \frac{30}{13}$$

$$n = 2: \qquad \frac{p_3}{q_3} = \frac{a_3 p_2 + p_1}{a_3 q_2 + q_1} = \frac{5 \cdot 30 + 7}{5 \cdot 13 + 3} = \frac{157}{68}$$

$$n = 3: \qquad \frac{p_4}{q_4} = \frac{a_4 p_3 + p_2}{a_4 q_3 + q_2} = \frac{6 \cdot 157 + 30}{6 \cdot 68 + 13} = \frac{972}{421}$$

The last convergent is equal to the value of the original continued fraction. In other words, we have

$$[2;3,4,5,6] = \frac{972}{421}$$

Let us also check that the property of the penultimate convergent claimed in Fact 17.11 holds for this example. Specifically, we check that $x = 68, y = 157$ is an integral solution of $972x - 421y = -1$. Since there are five partial quotients, an odd number, the right side of this equation is -1 instead of $+1$ which confirms Fact 17.11.

Every nonzero rational number can be expressed as a continued fraction in exactly two different ways. In one of the alternate representations, the last partial quotient is always equal to 1. For example, it is easy to check that we have

$$[2; 3, 4, 5, 6] = [2; 3, 4, 5, 5, 1]$$

Example 17.15. Use continued fractions to find integer solutions x and y to the equation

$$17x - 60y = 1 \tag{17.36}$$

Solution. Using the Euclidean algorithm, we compute

$$17 = 0 \cdot 60 + 17$$
$$60 = 3 \cdot 17 + 9$$
$$17 = 1 \cdot 9 + 8$$
$$9 = 1 \cdot 8 + 1$$
$$8 = 8 \cdot 1$$

The divisors are the partial quotients of the continued fraction

$$\frac{17}{60} = [0; 3, 1, 1, 8]$$

Since there are an odd number of partial quotients, Fact 17.11 asserts that the penultimate convergent of this continued fraction gives a solution of

$$17x - 60y = -1$$

Since equation (17.36) has $+1$ on the right side, we use $[0; 3, 1, 1, 7, 1]$, the alternate continued fraction representation of $17/60$ with an even number of partial quotients. The penultimate convergent of this continued fraction is $15/53$, which yields the solution $x = 53$, $y = 15$ of equation (17.36). By Fact 17.11, the general solution of equation (17.36) is

$$x = 53 + 60k \quad y = 15 + 17k$$

where k is an arbitrary integer.

Irrational Numbers

We have seen that an arbitrary rational number p/q can be represented as a continued fraction in exactly two different ways. Since the Euclidean algorithm ends in a finite number of steps, a rational number always has *finitely* many partial quotients. In this section, we will see that it is possible to define *infinite* continued fractions.

Fact 17.16. *Let $a_0, a_1, a_2, \ldots$, be an infinite sequence of integers, all of them positive except a_0, which may be zero. Then there exists a unique irrational positive number x such that the convergents (17.35) of the continued fraction $[a_0; a_1, a_2, a_3, \ldots]$ tend to the irrational number x. That is, any prescribed closeness of approximation is achieved by convergents p_n/q_n with n sufficiently large.*

Each irrational number has exactly one representation as a continued fraction, in contrast to the rational numbers, which have exactly two such representations.

Since arbitrary nonzero real numbers can be represented as continued fractions, they have been studied as a variant basis for the arithmetic of real numbers.[3] The other options for arithmetic are (1) rational numbers and (2) decimal fractions. The arithmetic of continued fractions is more complex than the arithmetic of decimal fractions—with the following exception. Computing reciprocals is just as easy for continued fractions as it is for rational numbers. For rational numbers the trivial computation of reciprocals is

$$\frac{1}{p/q} = \frac{q}{p}$$

and for continued fractions it is (assuming $a_0 > 0$)

$$\frac{1}{[a_0; a_1, a_2, \ldots]} = [0; a_0, a_1, a_2, \ldots]$$

There is no such easy formula for the computation of reciprocals of decimal fractions.

Continued fraction arithmetic is a computer alternative to the arithmetic of rational numbers for infinite precision arithmetic—calculation without roundoff error. Infinite precision arithmetic for rational numbers leads to huge numbers in computations of even modest scope. Decimal fractions are the preferred representation for most purposes.

The Fibonacci sequence

On page 48, we saw that the repeating decimals are identical with the rational numbers. There is also a characterization of the repeating continued fractions that we will state in Fact 17.20, but first let us see a few examples. We start with the simplest infinite continued fraction $[1; 1, 1, 1, \ldots]$. Using a notation similar to the one we used on page 49 for repeating decimals, we write this continued fraction as $[1; \overline{1}]$. Using equations (17.35) we obtain the convergents of this continued fraction.

$$\frac{p_0}{q_0} = 1 \qquad\qquad \frac{p_1}{q_1} = 2$$

$$\frac{p_2}{q_2} = \frac{3}{2} = 1.5 \qquad\qquad \frac{p_3}{q_3} = \frac{5}{3} = 1.\overline{6}$$

$$\frac{p_4}{q_4} = \frac{8}{5} = 1.6 \qquad\qquad \frac{p_5}{q_5} = \frac{13}{8} = 1.625$$

$$\frac{p_6}{q_6} = \frac{21}{13} \approx 1.61538 \qquad\qquad \frac{p_7}{q_7} = \frac{34}{21} \approx 1.61905$$

$$\frac{p_8}{q_8} = \frac{55}{34} \approx 1.61765 \qquad\qquad \frac{p_9}{q_9} = \frac{89}{55} = 1.6\overline{18}$$

$$\cdot \qquad \cdot \qquad \cdot$$

The numerators and denominators of the convergents are adjacent terms in the sequence:

$$1,\ 1,\ 2,\ 3,\ 5,\ 8,\ 13,\ 21,\ 34,\ 55,\ 89, \ldots$$

This sequence in which each term is the sum of the two previous terms is called the Fibonacci sequence after Leonardo Fibonacci (c. 1170–c. 1250) of Pisa, the greatest European mathematician of the Middle Ages. He mentioned this sequence in his book *Liber Abaci*, The Book of the Abacus, in connection with a problem about the growth of population of a colony of rabbits. Many phenomena of biological growth have been linked to the Fibonacci sequence. These include the spiral growth of the nautilus shell, the branching of plants, the number of petals on flowers, the double spiral arrangement of florets in a sunflower head and in pine cones, and the arrangement of leaves on a stem.

Fibonacci search

In computer science there is an algorithm called Fibonacci search. For example, suppose that we want use a computer to find a name N in a phone book.[4] We assume that the computer can immediately find a name with a certain *numerical* rank counting from the beginning of the phone book. If the number of names in the phone book is equal to the nth Fibonacci number F_n, then the search algorithm starts by finding the name M whose numerical rank is equal to F_{n-1}. Note that the number of names with rank greater than F_{n-1} is equal to F_{n-2} because from the definition of the Fibonacci sequence we have

$$F_n = F_{n-1} + F_{n-2}$$

1. *If $M = N$*, then we are done because we are searching for the name N, and we have found it.

2. *If the alphabetic rank of M is greater than N*, then we know that the name N is in the first part of the phone book. We repeat the above procedure using the

names from the beginning to M. Since there are F_{n-1} such names, we use the name L with the numerical rank F_{n-2} to divide the first half of the book into subsections of length F_{n-2} and F_{n-3}. Determine which subsection contains the name N by finding whether L is alphabetically greater or smaller than N.

Dangerous curve! ↤

3. *If the alphabetic rank of M is less than N,* then we know that the name N is in the last part of the phone book. We repeat the above procedure using the names from M to the end. We renumber the numerical ranking starting from the name following M. Since there are F_{n-2} names in the second part of the book, we use the name K with the numerical rank F_{n-3} to divide the second half of the book into subsections of length F_{n-3} and F_{n-4}. Determine which subsection contains the name N by finding whether K is alphabetically greater or smaller than N.

Quadratic equations can be solved by a process called *completing the square*. The quadratic equation

$$x^2 + x - 1 = 0$$

is equivalent to

$$x^2 + x + 1/4 = 5/4$$

This form of the equation now has the advantage that the left side is equal to a perfect square $(x+1/2)^2$. Taking square roots of both sides, we obtain

$$x = (1 \pm \sqrt{5})/2$$

Applying the method of completing the square to the general quadratic equation

$$ax^2 + bx + c = 0$$

one finds the roots

$$x = \frac{-b \pm \sqrt{b^2 - 4ac}}{2a}$$

This equation is known as the *quadratic formula*. (Recall that a solution of an equation is also called a *root* of the equation.)

Having carried out steps 2 or 3, we find that the name N is in a certain subsection whose length is again a Fibonacci number. We start the procedure from the beginning using this subsection instead of the entire book. Proceeding in this way, we eventually find the name we are searching for. The maximum number of steps in this process is n, a smaller number of steps than required by other methods.

The golden section

The convergents of the continued fraction $[1; \overline{1}]$ are quotients of Fibonacci numbers. This continued fraction $[1; \overline{1}]$ represents a particular irrational number g called the *golden section*.

Problem 17.17. Show that g is equal to $\frac{1+\sqrt{5}}{2}$.

Solution. Notice that $1/g$ is equal to $[0; \overline{1}]$ which is also equal to $g - 1$. From the equation $g - 1 = 1/g$, by multiplying both sides by g we obtain $g^2 + g - 1 = 0$, a quadratic equation with two roots:

$$g_1 = \frac{1 + \sqrt{5}}{2} \approx 1.61803$$

$$g_2 = \frac{1 - \sqrt{5}}{2} \approx -0.61803$$

Since we know that g is positive, it must be equal to the positive root, approximately 1.61803.

A   B   C

In the above diagram, the point B divides the line segment AC such that

$$\overline{AB}/\overline{BC} = g$$

The ancient Greeks first used the term *golden section* for this division of a line segment. For the Greeks, the significant property of this division is that the ratio of the larger segment to the whole, $\overline{AB} : \overline{AC}$, is equal to the ratio of the smaller to the larger segment, $\overline{BC} : \overline{AB}$. It is believed that the Pythagoreans considered g to be significant also because it is the ratio of the diagonal to the side of a regular pentagon.

The Greeks also believed that a rectangle with sides in the ratio of the golden section g has the most pleasing proportion. Here is an example of such a rectangle.

```
Golden
Section
Rectangle
```

The golden section has been used in the creations of architects and artists from ancient to modern times. The façade of the Parthenon of Athens (c. 450 B.C.) fits the golden section proportions almost exactly. In modern times, the architect Le Corbusier (1887–1965) used the golden section in the design of the U.N. building in New York. The French neoimpressionist artist Georges Seurat (1859–91) used many instances of the golden section in his painting *La Parade*.

Quadratic irrationalities

A number of the form $a + b\sqrt{D}$ is called a *quadratic irrationality* if a and b are rational numbers and D is a natural number that is not a perfect square. In Problem 17.17, we found the continued fraction representation of a particular quadratic irrationality, the golden section g.

We previously showed on page 48 that $\sqrt{2}$ is irrational. As an example, we will compute the continued fraction representation of $\sqrt{2}$, but first we need to recall a few facts from elementary algebra—in particular, the procedure known as rationalizing the denominator.[5] The following example will recall this method.

> Rationalizing the denominator makes it easier to compute decimal approximations by pencil and paper because one avoids division by awkward decimal fractions. However, this procedure is somewhat anachronistic today because it provides no advantage when computing with an electronic calculator.

Example 17.18. Rationalize the denominator of the following fraction:

$$\frac{1}{5 + 3\sqrt{2}}$$

Solution. Multiply the numerator and denominator by $5 - 3\sqrt{2}$:

$$\frac{1}{5 + 3\sqrt{2}} \cdot \frac{5 - 3\sqrt{2}}{5 - 3\sqrt{2}} = \frac{5 - 3\sqrt{2}}{5 \cdot 5 - 3 \cdot 3 \cdot \sqrt{2} \cdot \sqrt{2}} = \frac{5 - 3\sqrt{2}}{25 - 9 \cdot 2} = \frac{5 - 3\sqrt{2}}{7}$$

In the following, we use the reverse of this method; we rationalize the numerator, but the technique is almost identical. We also make use of the fact that we know that $\sqrt{2}$ is a number between 1 and 2.

Example 17.19. Find the continued fraction representation of $\sqrt{2}$.

Solution.

$$\sqrt{2} = 1 + (\sqrt{2} - 1) = 1 + \frac{1}{\sqrt{2} + 1}$$

$$= 1 + \frac{1}{2 + (\sqrt{2} - 1)} = 1 + \frac{1}{2 + \frac{1}{\sqrt{2}+1}}$$

$$= 1 + \frac{1}{2 + \frac{1}{2 + (\sqrt{2} - 1)}} = 1 + \frac{1}{2 + \frac{1}{2 + \frac{1}{\sqrt{2}+1}}}$$

At this point it can be seen that this is a repetitive process. The continued fraction representation of $\sqrt{2}$ is $[1; 2, 2, 2, \ldots]$—a repeating continued fraction that we denote $[1; \overline{2}]$.

Fact 17.20. *A quadratic irrationality, that is, a number of the form $a + b\sqrt{D}$ where a and b are rational numbers and D is a natural number that is not a perfect square, can be represented as a repeating continued fraction.*

Illustrating Fact 17.20, here are a few examples of repeating continued fractions for a few classes of irrational square roots.

$\sqrt{d^2 - 1}$	$\sqrt{d^2 + 1}$
$\sqrt{3} = [1; \overline{1, 2}]$	$\sqrt{2} = [1; \overline{2}]$
$\sqrt{8} = [2; \overline{1, 4}]$	$\sqrt{5} = [2; \overline{4}]$
$\sqrt{15} = [3; \overline{1, 6}]$	$\sqrt{10} = [3; \overline{6}]$
$\sqrt{24} = [4; \overline{1, 8}]$	$\sqrt{17} = [4; \overline{8}]$
$\sqrt{35} = [5; \overline{1, 10}]$	$\sqrt{26} = [5; \overline{10}]$

$\sqrt{d^2 + 2}$	$\sqrt{d^2 - 2}$
$\sqrt{3} = [1; \overline{1, 2}]$	$\sqrt{2} = [1; \overline{2}]$
$\sqrt{6} = [2; \overline{2, 4}]$	$\sqrt{7} = [2; \overline{1, 1, 1, 4}]$
$\sqrt{11} = [3; \overline{3, 6}]$	$\sqrt{14} = [3; \overline{1, 2, 1, 6}]$
$\sqrt{18} = [4; \overline{4, 8}]$	$\sqrt{23} = [4; \overline{1, 3, 1, 8}]$

Convergents as approximations

The continued fraction for the number π is $[3; 7, 15, 1, 292, 1, 1, \ldots]$. It is interesting to compare the first few convergents with the decimal expansion of π. (The reader can use equation (17.35) on page 247 to verify these numbers.)

$$3 = 3.00000000$$

$$\frac{22}{7} \approx 3.14285714$$

$$\frac{333}{106} \approx 3.14150943$$

$$\frac{355}{113} \approx 3.14159292$$

$$\pi \approx 3.14159265$$

Apart from $333/106$, all of the numbers in this list have been traditionally used as approximations to π. Notice that $355/113$ is correct to six decimal places. There is a general reason why convergents are such good approximations.

Fact 17.21. *Let p/q be one of the convergents of the continued fraction representation of a real number x. Then p/q approximates x better than any fraction whose denominator is no greater than q.*

Additionally, the convergent $355/113$ is an exceedingly good approximation for π because it happens to be the truncation of the continued fraction of π just before the large partial quotient 292.

The calendar

Astronomers tell us that the earth currently makes a circuit of the sun in 365.2422 days. This value, the tropical year, is slowly decreasing; it loses as much as .00052 days per year (about 45 seconds), a cumulative loss in 10,000 years of about 2.6 days. To synchronize the calendar with the tropical year length, most years should have 365 days, but some years, called leap years, have 366 days. There are several systems for achieving this: the Julian, Jalaali, and Gregorian calendars. Before discussing each of these systems, let us see how continued fractions can help with this problem. We start by expanding 0.2422, the portion of a day by which the tropical year exceeds 365 days, into the following continued fraction:

$$[0; 4, 7, 1, 3, 4, 1, 1, 1, 2] \tag{17.37}$$

This continued fraction terminates because 0.2422 is a rational number. Using equation (17.35), the convergents of (17.37) are found to be

$$\frac{p_0}{q_0} = 0$$

$$\frac{p_1}{q_1} = \frac{1}{4} \qquad\qquad = 0.25$$

$$\frac{p_2}{q_2} = \frac{7}{29} \qquad\qquad \approx 0.2413793103448$$

$$\frac{p_3}{q_3} = \frac{8}{33} \qquad\qquad \approx 0.2424242424242$$

$$\frac{p_4}{q_4} = \frac{31}{128} \qquad\qquad = 0.2421875$$

$$\frac{p_5}{q_5} = \frac{132}{545} \qquad\qquad \approx 0.2422018348624$$

$$\frac{p_6}{q_6} = \frac{163}{673} \qquad\qquad \approx 0.2421991084695$$

$$\frac{p_7}{q_7} = \frac{295}{1218} \qquad\qquad \approx 0.2422003284072$$

$$\frac{p_8}{q_8} = \frac{458}{1891} \qquad\qquad \approx 0.2421998942359$$

$$\frac{p_9}{q_9} = \frac{1211}{5000} \qquad\qquad = 0.2422$$

The above list confirms that the continued fraction (17.37) represents 0.2422 without any roundoff error. However, since p_4/q_4 already achieves four-figure agreement with 0.2422, the further convergents are not numerically significant.

Although Julius Cæsar was probably not acquainted with continued fractions, the *Julian calendar* uses the same approximation as the one given by the convergent $p_1/q_1 = 1/4$ by introducing a year 365 days long with a leap year every 4 years. The Julian calendar introduces an error at the rate of about one day per century.

The *Jalaali calendar*—which is official in Iran, Afghanistan, the Central Asian republics, and Kurdish Mesopotamia—was so named by the Iranian poet and mathematician Omar Khayyám (died c. 1123), who amplified that system. The Jalaali calendar specifies 8 leap years in every cycle of 33 years, resulting in a discrepancy with the tropical year at a rate of about 2 days every 10,000 years. This system uses the convergent $p_3/q_3 = 8/33$ as an approximation.

The *Gregorian calendar*—named after Pope Gregory XIII who introduced these changes in 1582—specifies a leap year every 4 years, except that years divisible by 100 *are not* leap years and years divisible by 400 *are* leap years. In a Gregorian cycle of 400 years there are 97 leap years. The Gregorian calendar is not based on a convergent of 0.2422; it uses instead the approximation

$$\frac{97}{400} = 0.2425$$

which leads to a discrepancy at the rate of about 3 days every 10,000 years. The comparison between the Jalaali and Gregorian calendars confirms Fact 17.21, which asserts that for the size of the numbers involved, the best rational approximations of a number are the convergents of the continued fraction expansion of that number.

A calendar for the Binary Era. The Jalaali calendar, based on p_3/q_3 is the most accurate of the three calendars discussed above. However, from Fact 17.21 it follows that $p_4/q_4 = 31/128$ must be more accurate still. In fact, p_4/q_4 leads to a calendar that is simpler than either the Jalaali or Gregorian calendars with an error rate of just a few hours in 10,000 years. Since the value of the tropical year is decreasing, it has the further advantage that the convergent p_4/q_4 is a slightly low estimate. The rule is this: *Years that are divisible by* 4 *are leap years except the years that are divisible by* 128. In other words, according to this rule, *a year is a leap year if its binary representation ends in at least two zeros but no more than six.*

In this chapter we have treated three major topics of number theory: diophantine equations, the Euclidean algorithm, and continued fractions.[6] In the next chapter, we will see that spies might be interested in finding very large prime numbers.

Chapter 18

Secret Messages

"And you really solved it?"

"Readily; I have solved others of an abstruseness ten thousand times greater. Circumstances, and a certain bias of mind, have led me to take interest in such riddles, and it may well be doubted whether human ingenuity can construct an enigma of the kind which human ingenuity may not, by proper application, resolve."

— EDGAR ALLAN POE, *The Gold-Bug*

When Winston Churchill said, "Never in the course of human events have so many owed so much to so few," he was referring to the airmen who died in the Battle of Britain. However, he could very well have added his own name to the list. Another name that might be added is that of mathematician Alan Turing (1912–54)—for the course of World War II might have been quite different if Turing had not broken the Enigma cipher, which Nazi Germany used to communicate with U-boats in the North Atlantic. Today cryptography is at the center of a debate over national security and individual liberty.

Some seek to suppress a secret, and that secret is the science of secrecy. Since the publication (Rivest, Shamir, and Adleman, 1978) of the discovery of the public-key cryptosystem known as RSA,[1] the U.S. National Security Agency has sought to control developments in this field.

The theory of RSA belongs to the field of mathematics known as *number theory*, the study of the properties of integers. It is an irony that number theory, once considered the purest of pure mathematics, now has applications that are so potent that government control is deemed necessary. We will discuss the mathematics of RSA; however, there are certain preparatory topics, interesting in their own right, that we must discuss first. These topics include Fermat's little theorem (not to be confused with Fermat's *last* theorem) and Euler's phi-function. In addition, we will make use of material introduced in Chapters 7 and 17.

Deciphering an RSA enciphered message is equivalent to factoring a number that happens to be the product of two very large prime numbers. Currently *very large* means about 160 digits. As the technology proceeds, this may be replaced

cryptosystem. A method of enciphering and deciphering messages.

plaintext. An unenciphered message.

ciphertext. An enciphered message.

Pronunciation. U.S. mathematicians and scientists generally pronounce the Greek letter phi (ϕ) as FEE. Most other speakers of English say FIE. This and other idiosyncratic pronunciations are probably due to two facts:
1. Before World War I, many prominent American mathematicians studied mathematics in Germany.
2. Starting about 1933, refugee mathematicians from continental Europe influenced a generation of American mathematicians.

with a larger number. Currently, it appears to be extremely difficult—practically impossible—to factor such a number in a reasonable length of time, even with the help of the fastest computers. However, so far no one knows for certain that there does not exist a method for accomplishing this task, and that is a source of anxiety in some quarters.

Fermat's Little Theorem

This and the following sections make use of modular arithmetic, a topic that was introduced in Chapter 7.

The relation of congruence (Definition 7.3) has many properties in common with equality, and this means that calculation with congruences is very much like ordinary arithmetic. The following fact summarizes some of the simplest properties of congruences. Verifying them would be a good review of modular arithmetic.

Fact 18.1. *Let n be a natural number, and let a, b, c, u, and v be integers, and let $a \equiv b$ (mod n).*

1. If $b \equiv c$ (mod n), then $a \equiv c$ (mod n).

2. If $c \equiv d$ (mod n), then $au + cv \equiv bu + dv$ and $ac \equiv bd$ (mod n).

3. If d divides n ($d \mid n$) and $d > 0$, then $a \equiv b$ (mod d).

Definition 18.2. Let n be a natural number and let x and y be integers. If $x \equiv y$ (mod n), then x is called a *residue of y modulo n*. In addition, if $0 \leq x < n$, then we write $x = y$ mod n and we say that x is the *least residue of y modulo n*.

If y is positive, to reduce y modulo n means to find the remainder after dividing y by n. Here is the general definition.

Definition 18.3. To *reduce y modulo n* means to find the least residue of y modulo n.

To reduce a large power of a number with respect to a modulus, it is very useful to find the binary representation of the exponent in order to use the method of Example 7.9.

Example 18.4. Reduce 9^{53} modulo 77.

Solution. The binary representation of 53 is 110101_2, that is, $53 = 32 + 16 + 4 + 1$. We reduce the powers of the form 9^{2^k} modulo 77 as follows:

$$9^2 = 81 \equiv 4 \qquad\qquad\qquad\qquad\qquad\text{(mod 77)}$$

$$9^4 = \left(9^2\right)^2 \equiv 4^2 = 16 \qquad\qquad\qquad\text{(mod 77)}$$

$$9^8 = \left(9^4\right)^2 \equiv 16^2 = 256 = 3 \cdot 77 + 25 \equiv 25 \qquad\text{(mod 77)}$$

$$9^{16} = \left(9^8\right)^2 \equiv 25^2 = 625 = 8 \cdot 77 + 9 \equiv 9 \qquad\text{(mod 77)}$$

$$9^{32} = \left(9^{16}\right)^2 \equiv 9^2 = 81 = 77 + 4 \equiv 4 \qquad\text{(mod 77)}$$

Notation. There are two different meanings to the symbol mod.

1. *When it appears in parentheses.* The expression

$$3 \equiv 45 \quad \text{(mod 6)}$$

is read, "3 is congruent to 45 modulo 6."

2. *Without parentheses.* The expression

$$3 = 45 \bmod 6$$

is read, "3 is equal to 45 reduced modulo 6."

Note that there are infinitely many integers congruent to 45 (mod 6). On the other hand, 45 mod 6 is equal to a particular number, the number 3.

The computer language Pascal defines mod for negative numbers in a manner that is inconsistent with standard mathematical usage. For example, in most implementations of Pascal -17 mod 10 represents -7 whereas in mathematics it is $+3$. In the C language, % is the same as mod in Pascal.

Now we compute 9^{53} as follows.

$$9^{53} = 9^{32+16+4+1} = 9^{32}9^{16}9^49^1 \equiv 4 \cdot 9 \cdot 16 \cdot 9 = 5184 = 67 \cdot 77 + 25 \equiv 25 \quad (\text{mod } 77)$$

In other words, $9^{53} = 25 \pmod{77}$.

Most of the arithmetic properties of congruences are like the corresponding properties of ordinary arithmetic, but cancellation is not quite as simple. For example, $3 \cdot 1 \equiv 3 \cdot 3 \pmod 6$, but we may not cancel 3 from both sides of this congruence because that would result in $1 \equiv 3 \pmod 6$, which is false. However, we are allowed to cancel under certain conditions as shown by the following proposition. (Recall the definition of *relatively prime*, Definition 16.2.)

Proposition 18.5. *Let the integer k be relatively prime to the natural number n. Then if $ka \equiv ka' \pmod n$ then $a \equiv a' \pmod n$.*

Proof. The assumption $ka \equiv ka' \pmod n$ means that $k(a - a')$ is divisible by n. Since k is relatively prime to n, it follows that $a - a'$ must be divisible by n. In other words, $a \equiv a' \pmod n$. □

The following fact was discovered by Fermat.

Theorem 18.6 (Fermat's little theorem). *Let p be a prime, and let a be an integer not a multiple of p. Then*

$$a^{p-1} \equiv 1 \quad (\text{mod } p) \tag{18.1}$$

Moreover, for every integer m we have

$$m^p \equiv m \quad (\text{mod } p) \tag{18.2}$$

For example, putting $p = 5$ and $a = 3$,

$$3^{5-1} = 81 \equiv 1 \quad (\text{mod } 5)$$

Proof. None of the numbers

$$a, \, 2a, \, 3a, \ldots, (p-1)a \tag{18.3}$$

is divisible by p. Moreover, no two of them can be congruent modulo p. In fact, suppose $qa \equiv ra \pmod p$ for some q and r $(q \neq r)$. Then, by Proposition 18.5,

$$q \equiv r \quad (\text{mod } p)$$

which is impossible. This implies that the numbers (18.3) must be congruent to the numbers $1, \, 2, \ldots, p-1$ possibly rearranged in a different order. This means that the product of the numbers (18.3) must be congruent to the product of the numbers $1, \, 2, \ldots, p-1$. That is,

$$1 \cdot 2 \cdot 3 \cdots (p-1) \equiv 1 \cdot 2 \cdot 3 \cdots (p-1)a^{p-1} \quad (\text{mod } p)$$

The number $1 \cdot 2 \cdot 3 \cdots (p-1)$ is not divisible by p and hence may be canceled from both sides of this congruence, which yields equation (18.1). This equation implies equation (18.2) unless m is divisible by p, but in this case equation (18.2) is clearly true. □

To test Theorem 18.6 with larger numbers, we use the technique of Example 18.4.

Example 18.7. Show $43^{60} \equiv 1 \pmod{61}$.

Solution. Since 61 is a prime, the result follows from Theorem 18.6. To confirm this by computation, we start by finding the binary representation of the exponent 60. Using the methods of Chapter 8, we have

$$60_{10} = 111100_2 = 32 + 16 + 8 + 4$$

Now reduce numbers of the form 43^{2^k} modulo 100 as follows.

$$43^2 = 1849 = 30 \cdot 61 + 19 \equiv 19 \qquad \pmod{61}$$
$$43^4 = \left(43^2\right)^2 \equiv 19^2 = 361 = 5 \cdot 61 + 56 \equiv 56 \qquad \pmod{61}$$
$$43^8 = \left(43^4\right)^2 \equiv 56^2 = 3136 = 51 \cdot 61 + 25 \equiv 25 \qquad \pmod{61}$$
$$43^{16} = \left(43^8\right)^2 \equiv 25^2 = 625 = 10 \cdot 61 + 15 \equiv 15 \qquad \pmod{61}$$
$$43^{32} = \left(43^{16}\right)^2 \equiv 15^2 = 225 = 3 \cdot 61 + 42 \equiv 42 \qquad \pmod{61}$$

Now we are ready to reduce 43^{60} modulo 61.

$$43^{60} = 43^{32+16+8+4} = 43^{32} \cdot 43^{16} \cdot 43^8 \cdot 43^4$$
$$\equiv 42 \cdot 15 \cdot 25 \cdot 56 = 882,000 \equiv 1 \qquad \pmod{61}$$

The last congruence can be verified by observing that division of $882,000$ by 61 leaves a remainder of 1. This confirms Theorem 18.6 for this particular example.

We can use Fermat's little theorem to reduce large powers of a number with respect to a prime modulus. The following proposition is of assistance for that purpose.

Proposition 18.8. *Let p be a prime and let a and s be integers. Then $a^s \equiv a^t$ (mod p) where t is a residue of s modulo $p - 1$.*

Proof. There exists an integer k such that $s = t + k(p - 1)$. From Fermat's little theorem, we have

$$a^s = a^{t+k(p-1)} = a^t \left(a^{p-1}\right)^k \equiv a^t \pmod{p} \qquad \square$$

In the following example, we use Proposition 18.8, putting t equal to the *least* residue modulo $p - 1$.

Example 18.9. Reduce 43^{1234} modulo 61; that is, find $43^{1234} \bmod 61$

Solution. First we reduce 1234 modulo 60:

$$1234 = 20 \cdot 60 + 34 \equiv 34 \pmod{60}$$

By Proposition 18.8, $43^{1234} \equiv 43^{34}$ (mod 61). Now we find the binary representation $34 = 100010_2 = 32 + 2$. Now use the calculations of Example 18.7 to obtain

$$43^{1234} \equiv 43^{34} = 43^{32} \cdot 43^2 \equiv 42 \cdot 19 = 798 = 13 \cdot 61 + 5 \equiv 5 \quad (\text{mod } 61)$$

In other words, we have $43^{1234} = 5$ mod 61.

The following is a consequence of Theorem 18.6 similar to Proposition 18.8.

Proposition 18.10. *Let p be a prime, and let q be an integer such that*

$$q \equiv 1 \quad (\text{mod } p - 1)$$

Then, for any integer m, we have

$$m^q \equiv m \quad (\text{mod } p)$$

Proof. Since $q \equiv 1$ (mod $p - 1$), there exists an integer k such that $q = 1 + k(p - 1)$. Therefore, using Theorem 18.6, we have

$$m^q = m^{1+k(p-1)} = m \cdot m^{k(p-1)} = m \cdot \left(m^{p-1}\right)^k \equiv m \quad (\text{mod } p - 1) \qquad \square$$

Public-Key Ciphers

We interrupt the exposition of number theory for a moment to explain the concept of a public-key cipher and why Fermat's little theorem might be relevant. Suppose that Ada, Ben, and Cal wish to exchange secret messages. If they use a conventional private-key cipher, then the three of them share certain secret information called a key. For example, if Ada wants to send a message to Ben, she enciphers her message using this key and an enciphering algorithm, and Ben deciphers the message by using the same key and a deciphering algorithm.

On the other hand, if they use a public-key cipher, then there are three different *public enciphering keys* for messages intended for each of the three—Ada, Ben, and Cal. Moreover, they have three different *private deciphering keys* that they share with no one—not even each other. No attempt is made to keep the enciphering keys secret. They can be made public—hence, the name *public-key cipher*—because it is impossible to infer the deciphering keys from the enciphering keys. The three— Ada, Ben, and Cal—are not concerned that outsiders may use their enciphering keys to send them false messages because they also have a technical method, which we will discuss later, of authenticating their signatures.

A one-to-one mapping is called a *trapdoor* if it is easy to compute the mapping but extremely difficult—essentially impossible—to compute the inverse of the mapping. A public-key enciphering algorithm is an example of a trapdoor because with knowledge of the enciphering key it is easy to encipher a message but extremely difficult or impossible to decipher the same message without also knowing another key called a deciphering key.

We will attempt to base a cryptosystem on Fermat's little theorem (Theorem 18.6). However, we will see that such a system is not suitable for public-key cryptography because its enciphering algorithm is *not* a trapdoor. This cryptosystem is constructed as follows. In Proposition 18.10, put q equal to a product of two integers, an enciphering key e and a deciphering key d. Proposition 18.10 then asserts

$$m^{ed} \equiv m \pmod{p}$$

provided that $ed \equiv 1 \pmod{p-1}$. We suppose that the integer m $(0 < m < p)$ represents the plaintext message that we wish to encipher. (There are various ways of representing a message as an integer. For example, we could employ the numerical code used by computers, the ASCII code, to represent letters and characters.) The enciphering algorithm is

$$M = m^e \bmod p \tag{18.4}$$

where the ciphertext is the integer M. The deciphering algorithm is

$$m = M^d \bmod p \tag{18.5}$$

Using the fact that $ed \equiv 1 \pmod{p-1}$, this formula follows from Proposition 18.10. In fact, we have:

$$M^d \equiv (m^e)^d = m^{ed} \equiv m \pmod{p}$$

Example 18.11. Use the foregoing method to encipher the plaintext message $m = 43$, using the prime modulus $p = 61$ and the enciphering key $e = 17$. Find the ciphertext M and the deciphering key d. Demonstrate the deciphering algorithm.

Solution. We find the ciphertext using equation (18.4) as follows.

$$M = m^e \bmod p = 43^{17} \bmod 61$$

We now use the calculations of Example 18.7 to obtain $M = 35$ as follows.

$$43^{17} = 43^{16+1} = 43^{16} \cdot 43 \equiv 15 \cdot 43 = 645 = 10 \cdot 61 + 35 \equiv 35 \pmod{61}$$

Before we can decipher $M = 35$, we must find the deciphering code d. For e equal to any integer relatively prime to $p - 1 = 60$, it is possible to find d such that $ed = 1 \bmod (p-1)$; in our particular case we wish to find d such that $17d = 1 \bmod 60$. This problem is equivalent to solving the diophantine equation $17d - 60k = 1$ for the integers d and k. This problem was solved in Example 17.15, where we obtained $d = 53$ and $k = 15$. Therefore, with respect to the modulus $n = 61$, the deciphering key $d = 53$ belongs to the enciphering key $e = 17$. To decipher $M = 35$ we must compute

$$M^d \bmod p = 35^{53} \bmod 61$$

We use the technique of Example 18.7. First we reduce the powers of 35 modulo 61 as follows.

$$35^2 = 1225 = 20 \cdot 61 + 5 \equiv 5 \qquad (\text{mod } 61)$$

$$35^4 = \left(35^2\right)^2 \equiv 5^2 = 25 \qquad (\text{mod } 61)$$

$$35^8 = \left(35^4\right)^2 \equiv 25^2 = 625 = 10 \cdot 61 + 15 \equiv 15 \qquad (\text{mod } 61)$$

$$35^{16} = \left(35^8\right)^2 \equiv 15^2 = 225 = 3 \cdot 61 + 42 \equiv 42 \qquad (\text{mod } 61)$$

$$35^{32} = \left(35^{16}\right)^2 \equiv 42^2 = 1764 = 28 \cdot 61 + 56 \equiv 56 \qquad (\text{mod } 61)$$

Next we find the binary representation $53_{10} = 110101_2$; and finally we compute

$$35^{53} = 35^{32+16+4+1} = 35^{32} \cdot 35^{16} \cdot 35^4 \cdot 35$$

$$\equiv 56 \cdot 42 \cdot 25 \cdot 35 = 2,058,000 = 33,737 \cdot 61 + 43 \equiv 43 \qquad (\text{mod } 61)$$

This confirms that the ciphertext $M = 35$ deciphers back to the plaintext $m = 43$.

We have verified that deciphering the enciphered message yields the original plaintext as claimed. In the above problem, we used the small prime number 61. In practice, we might use a very large prime for p; nevertheless, such a cryptosystem has a very serious flaw. The enciphering key e and the modulus p are known to the public, and we have just seen that with this information it is possible to compute the deciphering key d. In other words, the enciphering algorithm is not a trapdoor. With the knowledge of the public enciphering key and the modulus, an unintended person can decipher messages.

We see that Fermat's little theorem does not succeed as a basis for a cryptosystem. Nevertheless, the RSA cryptosystem is a modification of the above method based on a generalization of Fermat's little theorem that we will discuss in the next section.

Euler's Generalization

Definition 18.12. Let n be a natural number. Then $\phi(n)$ denotes the number of natural numbers less than n that are relatively prime to n.

The phi-function

The numbers less than 12 that are relatively prime to 12 are:

$$1, \, 5, \, 7, \, 11$$

Therefore $\phi(12) = 4$.

There is a general formula for $\phi(n)$, but the following propositions give all that we need to know about ϕ to discuss RSA. In particular, we will show how to compute $\phi(n)$ in case n is a prime or a product of two different primes. The case in which n is a prime is easy because a prime number is relatively prime to all natural numbers smaller than n. Since there are $n-1$ such numbers, we have the following.

Proposition 18.13. *If p is a prime, then* $\phi(p) = p - 1$.

The product of two primes is just a little more difficult.

Proposition 18.14. *If n is the product of two distinct prime numbers p and q, that is, if n = pq, then*

$$\phi(n) = (p-1)(q-1)$$

Proof. The numbers less than n that are relatively prime to pq are precisely the numbers that do not contain p or q as a factor. This includes all the numbers less than pq (there are $pq - 1$ of them) except for the $p - 1$ numbers

$$q, \ 2q, \ 3q, \ldots, \ (p-1)q$$

that are divisible by q, and the $q - 1$ numbers

$$p, \ 2p, \ 3p, \ldots, \ (q-1)p$$

that are divisible by p. That is,

$$\phi(n) = (pq-1) - (p-1) - (q-1) = pq - p - q + 1 = (p-1)(q-1) \qquad \square$$

Residue classes and systems

The division algorithm (Theorem 8.2) asserts the following. Given any integer a, let q and r be the quotient and remainder when a is divided by a natural number n, that is,

$$a = qn + r$$

so that $a \equiv r \pmod{n}$. Since $0 \leq r < n$, every integer is congruent, modulo n to one of the numbers

$$0, \ 1, \ 2, \ldots, \ n-1 \tag{18.6}$$

Also, it is clear that no two of these numbers are congruent, modulo n. Indeed, if two of these numbers were congruent, modulo n, then the difference of these two numbers would be an integral multiple of n; but this is impossible because the difference between the larger and the smaller of these two numbers must be a positive integer less than n. The set of numbers (18.6) constitutes an example of what is called a complete residue system, modulo n. More generally, we define a complete residue system as follows.

Definition 18.15. A set of n integers

$$x_1, \ x_2, \ldots, \ x_n$$

is called a *complete residue system modulo n* if no two of the numbers are congruent modulo n.

Problem 18.16. Does 22, 11, 12, 21 constitute a complete residue system, modulo 4?

Solution. We need to check that 22 is not congruent to 11 (mod 4), 22 is not congruent to 12 (mod 4), and so on. Since no two of these four integers are congruent to each other, this set constitutes a complete residue system modulo 4.

An easier way to determine whether these four numbers constitute a complete residue system modulo 4 is to reduce each of these four numbers modulo 4 and obtain 2, 3, 0, 1 and to note that these are exactly the four possible remainders obtained when dividing integers by 4.

Definition 18.17. For a given integer a and a modulus $n > 0$, the set of all integers x satisfying

$$x \equiv a \quad (\text{mod } n)$$

is called a *residue class modulo n.*

The residue class determined by $x \equiv a$ (mod n) is the set of integers

$$\ldots, a - 3n, a - 2n, a - n, a, a + n, a + 2n, a + 3n, \ldots$$

A complete residue system, modulo n, therefore, consists of one representative from each of the n residue classes with respect to the modulus n.

Thus, there are six residue classes, modulo 6, of which the one determined by

$$x \equiv 5 \quad (\text{mod } 6)$$

consists of the numbers

$$\ldots, -13, -7, -1, 5, 11, 17, \ldots$$

Note that all numbers in this particular residue class are relatively prime to 6. This suggests that, if one integer in a residue class, modulo n, is relatively prime to n, all integers in that residue class, modulo n, are relatively prime to n. The following proposition states this fact.

Proposition 18.18. *If $b \equiv c$ (mod n) and b is relatively prime to n, then c is relatively prime to n.*

Proof.

1. The assumption $b \equiv c$ (mod n) means that there exists an integer k such that $b - c = kn$.

2. The assumption that b is relatively prime to n means that (b, n), the greatest common divisor (g.c.d.) of b and n, is equal to 1.

From 1 it follows that every common divisor of c and n must also be a divisor of b. It follows that (c, n), the g.c.d. of c and n, is a divisor of (b, n). Since according to 2, (b, n) is equal to 1, it follows that (c, n) is equal to 1, which means that c and n are relatively prime. □

It follows that, with respect to the modulus n:

- Each of the $\phi(n)$ natural numbers less than n and relatively prime to n belong to $\phi(n)$ different residue classes.

- The integers belonging to any one of these residue classes are all relatively prime to n.

- Every integer relatively prime to n belongs to exactly one of these residue classes.

- A set of $\phi(n)$ mutually incongruent integers that are relatively prime to n must contain exactly one representative from each of these residue classes.

Definition 18.19. A set of $\phi(n)$ mutually incongruent integers,

$$r_1, r_2, \ldots, r_{\phi(n)}$$

each of which is relatively prime to n, is called a *reduced residue system modulo n.*

Note that any reduced residue system modulo n can be enlarged to form a complete residue system. For example, the integers 1, 3, 7, 9 constitute a reduced residue system modulo 10 that can be enlarged to the complete residue system 0, 1, 2, ..., 10. Furthermore, we can create a reduced residue system modulo n by selecting the integers relatively prime to n from a complete residue system.

Proposition 18.20. *Let*

$$r_1, r_2, \ldots, r_k \tag{18.7}$$

be a complete $(k = n)$, or a reduced $(k = \phi(n))$, residue system modulo n, and let a be an integer relatively prime to n. Then

$$ar_1, ar_2, \ldots, ar_k \tag{18.8}$$

is also, respectively, a complete, or a reduced residue system, modulo n.

Proof. In either case (complete or reduced), there are the same number of integers in (18.8) and (18.7). Furthermore, no two of the integers in (18.8) can be congruent modulo n. Indeed, if $ar_i \equiv ar_j \pmod{n}$ and $i \neq j$, then by Proposition 18.5, we would have $r_1 \equiv r_j \pmod{n}$, which is impossible. Moreover, if r is relatively prime to n, then so is ar. The assertion now follows from Definitions 18.15 and 18.19. $\square$

Euler's theorem

Now we come to Euler's generalization of Fermat's little theorem.

Theorem 18.21. *Let n be a natural number greater than 1 and let a be an integer relatively prime to n. Then*

$$a^{\phi(n)} \equiv 1 \pmod{n}$$

Proof. Let

$$r_1, r_2, \ldots, r_{\phi(n)}$$

be a reduced residue system modulo n. Since

$$ar_1, ar_2, \ldots, ar_{\phi(n)}$$

is also a reduced residue system modulo n, it follows that the integers in the latter set must be congruent to some rearrangement of the integers in the former set. Therefore, the following congruence is true.

$$a^{\phi(n)} r_1 \cdot r_2 \cdots r_{\phi(n)} \equiv r_1 \cdot r_2 \cdots r_{\phi(n)} \pmod{n}$$

From Proposition 18.5 we have

$$a^{\phi(n)} \equiv 1 \pmod{n} \qquad \square$$

With the help of Theorem 18.21, we have a generalization of Proposition 18.8 that enables us to use a technique for reducing a large power with respect to a modulus.

Proposition 18.22. *Let n be an integer and let a be relatively prime to n. Further, let $s \equiv t \pmod{\phi(n)}$. Then*

$$a^s \equiv a^t \pmod{n}$$

Proof. There exists an integer k such that $s = t + k\phi(n)$. Therefore, using the fact that $a^{\phi(n)} \equiv 1 \pmod{n}$, we have

$$a^s = a^{t+k\phi(n)} = a^t a^{k\phi(n)} = a^t \left(a^{\phi(n)}\right)^k \equiv a^t \pmod{n} \qquad \square$$

The following example is a computational illustration of Proposition 18.22.

Example 18.23. Reduce 25^{1234} modulo 77.

Solution. Since 77 is a product of the two primes 7 and 11, by Proposition 18.14

$$\phi(77) = (7-1)(11-1) = 6 \cdot 10 = 60$$

Reducing 1234 modulo 60, we obtain 34. Since 77 and 25 are relatively prime, it follows from Proposition 18.22 that the original problem is the same as reducing 25^{34} modulo 77. Now we use the technique of Example 18.7. We reduce the powers of the form 25^{2^k} modulo 77.

$$25^2 = 625 \equiv 9 \qquad\qquad\qquad\qquad\qquad\qquad (\bmod\ 77)$$
$$25^4 = \left(25^2\right)^2 \equiv 9^2 = 81 = 77 + 4 \equiv 4 \qquad (\bmod\ 77)$$
$$25^8 = \left(25^4\right)^2 \equiv 4^2 = 16 \qquad\qquad\qquad\quad (\bmod\ 77)$$
$$25^{16} = \left(25^8\right)^2 \equiv 16^2 = 256 = 3 \cdot 77 + 25 \equiv 25 \quad (\bmod\ 77)$$
$$25^{32} = \left(25^{16}\right)^2 \equiv 25^2 = 625 = 8 \cdot 77 + 9 \equiv 9 \quad (\bmod\ 77)$$

The binary representation of 34 is $32 + 2$. Hence, we have

$$25^{1234} \equiv 25^{34} = 25^{32+2} = 25^{32} \cdot 25^2 \equiv 9 \cdot 9 = 81 = 77 + 4 \equiv 4 \pmod{77}$$

The RSA Cryptosystem

Previously, we saw that it is not possible to define an effective public-cipher crypto-system using Fermat's little theorem—using equations (18.4) and (18.5) with p equal to a prime number—because it is easy to infer the deciphering key from the knowledge of the corresponding enciphering key and the modulus. However, Euler's generalization of Fermat's little theorem gives us just the tools we need.

We base our public-key cipher not on a prime modulus but rather a modulus $n = pq$ that is a product of two distinct prime factors p and q. We must find e and d, both greater than 1, such that for an arbitrary message m we have $m^{ed} \equiv m \pmod{n}$. According to Theorem 18.21, we could choose e and d such that

$$ed = \phi(n) + 1 = (p-1)(q-1) + 1$$

but according to Proposition 18.22, it is sufficient to find e and d such that $ed \equiv 1 \pmod{\phi(n)}$. As before, we put e equal to an arbitrary number relatively prime to $\phi(n) = (p-1)(q-1)$. We must find integers d and k that satisfy the following equation.

$$de - k\phi(n) = 1 \tag{18.9}$$

We can solve this equation using the methods of Chapter 17.

In a moment we will exhibit a computational example of this procedure, but first let us discuss the following crucial question. *Is it possible to infer the deciphering key d from a knowledge of the enciphering key e and the modulus n = pq?* It is not currently possible if the prime numbers p and q, the factors of n, are sufficiently large. In fact, to break the cipher, we must find the deciphering key d based on the knowledge of the enciphering key e and the modulus n. We can do this by solving equation (18.9), but to do so, we need to determine $\phi(n)$, which is equal to $(p-1)(q-1)$, from the knowledge of n. It would be sufficient to find the prime factors of n. The fact that makes this algorithm a trapdoor is that it is extremely difficult to factor a number that is the product of two very large prime numbers. Currently, it is sufficient to use primes of more than 160 digits, but larger primes may be needed as algorithms for factoring large numbers improve.

> *Q. If these computations are done on electronic computers that can perform millions of operations in a second, why isn't it possible to factor these numbers?*
>
> *A. Even electronic computers have limitations. The following calculation may increase your appreciation of the magnitude of the difficulty of factoring large numbers. There are more efficient methods of factorization, but the straightforward method still gives a rough estimate of the difficulty. In this method, we test as possible divisors of n all odd numbers less than the square root of n. By this method, we eventually find all the factors of n. If p and q have 160 digits each, then n has an order of magnitude of 10^{320}. Merely to write all the digits of a number this large would require more than four full lines of print. To factor this*

*number by the straightforward method, we must carry out about 10^{160}
(the square root of 10^{320}) trial divisions. If we can do this on an electronic computer at the rate of one million every second, the task will
take about 10^{154} seconds. Since the number of seconds in a year is*

$$60 \cdot 60 \cdot 24 \cdot 365 \approx 3 \cdot 10^7$$

*it follows that in a year it is only possible to test about $3 \cdot 10^{13}$ numbers.
Therefore, the straightforward method will require about $3 \cdot 10^{140}$ years
to factor just one number. For comparison, the "Big Bang," the birth
of the universe, is said to have occurred about 10^{10} (10 billion) years
ago. In actuality, there are methods of factorization that are much better
than this straightforward method, but, nevertheless, the huge numbers
that we are discussing are beyond the capabilities of current methods.
It is conceivable that algorithms may be developed that can factor such
large numbers with sufficient speed; the users of RSA cryptosystems are
betting that this doesn't happen.*

*Q. Your calculation seems to show that it is also very time consum-
ing to test large numbers for primality. How is it possible to obtain a
sufficient supply of very large prime numbers?*

*A. Indeed, the straightforward method would take a very long time
to check the primality of a large number. However, there are other
methods, not based on factorization, for checking primality that are fast
enough to check 160-digit numbers in a reasonable length of time. RSA
cryptography remains effective so long as the fastest current algorithms
for factoring large numbers are very much slower than the current algorithms for checking primality.*

We will now illustrate the computational details for a small-scale RSA cryptosystem. For the modulus we will use 77, which is equal to the product of the primes 7
and 11. Of course, such a small modulus is completely inappropriate for an actual
cryptosystem because it is trivial to find the prime factors of 77, but we will use it to
keep the computation small. Since

$$\phi(77) = (7-1) \cdot (11-1) = 6 \cdot 10 = 60$$

e and d must be chosen such that $ed \equiv 1 \pmod{60}$. We can give e an arbitrary
value that is relatively prime to 60—let us use $e = 17$. Now d must be chosen such
that $17d \equiv 1 \pmod{60}$. It is an equivalent problem to find integers d and k such
that $17d - 60k = 1$. This is the very same equation that occurred in Example 18.11.
This problem was solved in Example 17.15 where we obtained $d = 53$ and $k = 15$.[2]
Note that $d = 53$ is the deciphering key, and recall that $e = 17$ is the enciphering key
relative to the modulus $n = 77$.

Example 18.24. Use $n = 77$, $e = 17$ to encipher the word *Yes* abbreviated as *Y*. That
is, since Y is the twenty-fifth letter of the alphabet, encipher the number 25.

Solution. Since $m^e = 25^{17}$, we reduce this number modulo 77. Using the methods of Example 18.23 we have

$$25^{17} = 25^{16+1} = 25^{16} \cdot 25^1 \equiv 25 \cdot 25 = 625 = 8 \cdot 77 + 9 \equiv 9 \pmod{77}$$

The enciphered message is the number 9.

Example 18.25. Use $n = 77$, $d = 53$ to decipher the message consisting of the number 9.

Solution. To decipher $M = 9$ using the deciphering key 53, we must reduce $M^e = 9^{53}$ modulo 77. In Example 18.4 we calculated $25 = 9^{53}$ mod 77. Therefore, the deciphered message is Y, the twenty-fifty letter of alphabet.

Examples 18.24 and 18.25 exhibit the principles of public-key cryptosystems. Of course, this particular cryptosystem would be very easy to break because the modulus 77 is easily factored into $7 \cdot 11$. A real cryptosystem uses a modulus that is the product of two prime numbers of at least 160 digits each. The security of such a system is based on the belief that it is virtually impossible to factor a product of two very large primes. It would add greatly to our confidence in using such a system if it could be proved rigorously that such factorizations are very difficult, but to date no one has done so.

> *Q. The assumption $m < n$, that the message is numerically less than the modulus, seems too restrictive. What happens if Ada wants to send a message to Ben that is larger than n?*
>
> *A. Ada, Ben, and Cal have agreed to divide their messages into blocks of a certain fixed size l. The three moduli must be large enough to encipher a block of this size.*
>
> *Q. Must Ada, Ben, and Cal use keys with different moduli?*
>
> *A. They must do so if they wish to keep their deciphering keys secret from each other.*[3]

Authentication

In using a public-key cipher there is a problem that does not occur with private-key ciphers. Since the enciphering key is public, how can Ada be sure that the message she receives that claims to be from Ben is really from Ben? To solve this problem, Ben sends to Ada an enciphered signature S in addition to the enciphered message M. Of course, he cannot use the same enciphering method for the signature that he uses for the message because then anyone could encipher a forged signature using Ada's public enciphering key.

For Ada and Ben there are four different keys. There are the public enciphering keys e_A and e_B for messages directed to Ada and Ben, respectively. In addition, there is the deciphering key d_A that only Ada knows, and a different deciphering key d_B that only Ben knows. Moreover, there are the public moduli n_A and n_B for the keys of Ada and Ben, respectively. Suppose m and s are Ben's plaintext message and signature, respectively. As we discussed, the ciphertext message M is equal to $m^{e_A} \bmod n_A$ that Ada deciphers using $m = M^{d_A} \bmod n_A$.

Ada surmises that one block of Ben's message is in fact his signature because the ordinary deciphering procedure yields gibberish.

There are two methods for constructing Ben's ciphertext signature S from his plaintext s depending on whether $n_B < n_A$ or $n_B > n_A$, and there are correspondingly two deciphering techniques for Ada. Both Ada and Ben know which of these inequalities is true because the moduli are public. (As previously discussed, it is not permitted that $n_B = n_A$.) Both methods have the novel feature that Ben uses his deciphering key as though it were an enciphering key, and Ada uses Ben's enciphering key as though it were a deciphering key. Depending on the relative sizes of the moduli, just one of the following methods is used.

1. $\mathbf{n_B} < \mathbf{n_A}$. Ben sends the ciphertext signature

$$S = \left(s^{d_B} \bmod n_B\right)^{e_A} \bmod n_A$$

The inequality $n_B < n_A$ ensures that the expression in parentheses is not too large to be enciphered by Ada's enciphering key.

Ada deciphers this using

$$s = \left(S^{d_A} \bmod n_A\right)^{e_B} \bmod n_B$$

2. $\mathbf{n_B} > \mathbf{n_A}$. Ben sends the ciphertext signature

$$S = \left(s^{e_A} \bmod n_A\right)^{d_B} \bmod n_B$$

The inequality $n_B > n_A$ ensures that the expression in parentheses is not too large to be enciphered by Ben's deciphering key.

Ada deciphers this using

$$s = \left(S^{e_B} \bmod n_B\right)^{d_A} \bmod n_A$$

As a result of this procedure, Ada has authenticated Ben's signature; she knows that the message really came from Ben.

In this chapter, we have seen that number theory is useful in cryptography.[4] In the next chapter, we will learn how mathematics can improve the operation of Cal's drapery factory.

Part V

WINNING

Chapter 19

Be Wise, Optimize

The good is the enemy of the best.

—Anonymous

In 1939, the Cowles Commission for Research in Economics began its work at the University of Chicago. It was charged with researching economics in a more mathematical way than had yet been attempted. The Cowles Commission was the prime mover in the development of the new science of econometrics, the application of mathematical methods to problems of economics. Their work continued through World War II because it appeared that problems of military strategy and tactics are similar to problems of economics. The new sciences of mathematical programming and operations research were born. As we will see, these new sciences occasionally achieved mathematical elegance.

> **Mathematical programming** must not be confused with **computer programming**. In mathematical programming, *to program* means *to schedule activities*. Computer programming could be used to solve a mathematical programming problem, but the two disciplines are otherwise unrelated.

Mathematical programming is concerned with problems of scheduling activities. The oldest and most elegant method of mathematical programming is called *linear programming*.

Linear Programming Examples

The diet problem

In 1947, the mathematical optimization technique known as *linear programming* became a practical tool for solving a wide variety of industrial and economic problems. In that year, the *simplex method*, which makes it computationally feasible to solve large-scale linear programming problems, was discovered by George B. Dantzig.[1] The new technique was soon applied to the *diet problem* that was earlier studied by Stigler in 1945.[2] The diet problem[3] seeks a menu that satisfies the requirements of human nutrition for the least cost. The five nutrients and nine foods considered in this historically important first application of the diet problem are shown in Figure 19.1.

To set up the problem quantitatively, we need to know the following three things.

Nutrients	Foods
1. Calories	1. Enriched wheat flour
2. Calcium	2. Evaporated milk
3. Vitamin A	3. Cheddar cheese
4. Riboflavin	4. Beef liver
5. Ascorbic acid	5. Cabbage
	6. Spinach
	7. Sweet potatoes
	8. Dried lima beans
	9. Dried navy beans

Figure 19.1. The five nutrients and nine foods for the minimum cost diet. The linear programming solution showed that, in 1947, the nutritional requirements of one person could be met using the above foods at a yearly cost of $39.69. These particular foods were chosen because they were considered cheap, nutritious, widely available, and not subject to seasonal fluctuations.

- The minimum daily requirement r_i of the ith nutrient. $(i = 1, \ldots, 5)$

- The unit cost c_j of the jth food. $(j = 1, \ldots, 9)$

- The concentration p_{ij} of the ith nutrient in the jth food.

To solve the problem, we must find daily amounts x_j $(j = 1, \ldots, 9)$ of each of the nine foods that satisfy each of the five nutritional requirements at minimal cost. The amount of the ith nutrient $(i = 1, \ldots, 5)$ in an amount x_j of the jth food is equal to $p_{ij}x_j$. The total amount of the ith nutrient from all nine foods is

$$p_{i1}x_1 + p_{i2}x_2 + \cdots + p_{i9}x_9 \ (1 \leq i \leq 5)$$

These total amounts of all five nutrients must be no less than r_i, the minimum daily requirement for the ith nutrient. That is, we must have

$$p_{i1}x_1 + p_{i2}x_2 + \cdots + p_{i9}x_9 \geq r_i \ (1 \leq i \leq 5)$$

Subject to the constraint of these five inequalities, we want to minimize the total cost. The cost of amount x_j of the jth food is c_jx_j. Therefore, the total cost is

$$c_1x_1 + c_2x_2 + c_3x_3 + \cdots + c_9x_9$$

The diet problem is an example of a linear programming problem. As noted in Figure 17.1, the graph consisting of points (x, y) that satisfy an equation such as $3x + 4y = 5$ is a straight line. This geometric fact justifies the following definition.

Definition 19.1. An expression of the form

$$a_1x_1 + a_2x_2 + \cdots + a_nx_n$$

is said to be *linear* in the variables $x_1, x_2, \ldots, x_n$.

With respect to the variables $x_1, x_2, \ldots, x_n$, the equation

$$a_1 x_1 + a_2 x_2 + \cdots + a_n x_n = b$$

is called a linear equation, and the inequalities

$$a_1 x_1 + a_2 x_2 + \cdots + a_n x_n \leq b$$

or

$$a_1 x_1 + a_2 x_2 + \cdots + a_n x_n \geq b$$

are called linear inequalities. We have seen that the diet problem leads to the problem of minimizing a linear expression subject to the condition that a system of linear inequalities is satisfied. The diet problem is expressed more generally in Figure 19.2.

> *Given costs c_j of N foods, daily requirements r_i for M nutrients, and concentrations p_{ij} of the ith nutrient in the jth food, find a menu of minimum cost such that, for each nutrient, the daily total of that nutrient is not less than the daily requirement. In other words, find N nonnegative food amounts x_j $(j = 1, 2, \ldots, N)$ such that*
>
> $$p_{i1} x_1 + p_{i2} x_2 + \cdots + p_{iN} x_N \geq r_i, \ (i = 1, 2, \ldots, M)$$
>
> *and such that the total cost*
>
> $$c_1 x_1 + c_2 x_2 + \cdots + c_N x_N$$
>
> *is minimum.*

Figure 19.2. The diet problem

Definition 19.2. A problem to determine nonnegative values of variables that minimize or maximize a linear expression (the *objective function*) subject to linear inequalities and/or equations (the *constraints*) is called a *linear programming problem*.

The diet problem is one of many scheduling problems that can be expressed as linear programming problems.

Geometric interpretation

In practical problems, such as the diet problem, the number of variables tends to be rather large. However, a linear programming problem involving just two variables has a geometric interpretation. We will consider an example.

Example 19.3. Find a point (x, y) in two-dimensional space such that:

1. Both x and y are nonnegative.

2. The following constraints are satisfied.

$$\begin{array}{ll} l_1: & 4x+y \geq 90 \\ l_2: & x+6y \geq 80 \\ l_3: & x+y \leq 60 \end{array} \qquad (19.1)$$

3. The objective function $z = 2x + y$ is minimized.

This problem is solved geometrically in Figure 19.3. (See the caption for details.) The minimum $z = 50$ is achieved at $x = 20$, $y = 10$. This problem can be solved as follows.

1. Find the points A, B, and C that are the vertexes of the constraint set. In this example, we have A $= (10, 50)$, B $= (56, 4)$, and C $= (20, 10)$.

2. Find the value of the objective function at each of these points. In this example, we have $z(A) = 70$, $z(B) = 116$, and $z(C) = 50$.

3. Since the smallest of these three values is 50, which is achieved at C, this is the solution to the problem.

> **convex.** A geometric figure is said to be *convex* if every two points of the set can be joined by a line segment that lies entirely in the set.

This geometric interpretation helps us understand the meaning of linear programming. In a two-dimensional problem with more than three constraints, the constraint set can be a convex polygon. In three dimensions, the constraint set becomes a convex polyhedron, and, in higher dimensions, a generalization of a polyhedron known as a *convex polytope*.

We continue with examples of scheduling problems that lead to linear programming problems.

The activity-analysis problem

The following linear programming problem concerns a factory that uses M resources (labor, machines, materials) to produce N different products. It is known that a_{ij} units of the ith resource are needed to produce one unit of the jth product. The amount available of the ith resource is b_i. Furthermore, c_j is the profit from producing one unit of the jth product. The manager wants to know how much of each product to produce to maximize profit. Specifically, she wants to schedule the optimal number x_j of the jth product. The algebraic representation of this problem is shown in Figure 19.4.

The transportation problem

The transportation problem was first stated by Hitchcock in 1941.[4] Fixed amounts of a commodity are available at several factories. Several warehouses each require specified amounts of the commodity. The unit costs are known for shipping the commodity from each factory to each warehouse. It is assumed that the total demand of

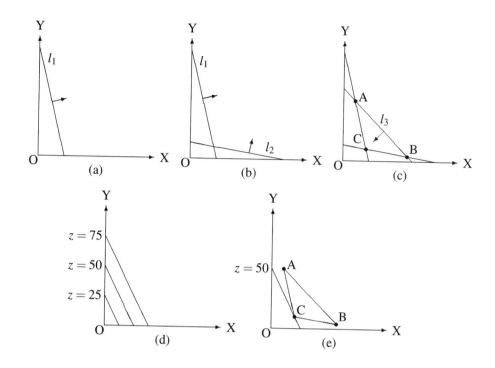

Figure 19.3. Graphical solution of Example 19.3. We only need to look at the first quadrant of the (x,y)-plane because a linear programming problem always requires that the variables are nonnegative. In (a), we see that the constraint l_1 in (19.1) requires that the point (x,y) be in the half plane, as shown by the arrow, to the right of the line $4x+y=90$. In (b), we add the constraint l_2 which requires that (x,y) be in the half plane above the line $x+6y=80$. In (c), we add the constraint l_3 which requires that (x,y) be below and to the left of the line $x+y=60$. Thus, the three constraints require that the point (x,y) be in or on the boundary of the triangle ABC. The coordinates of A, B, and C are $(10,50)$, $(56,4)$, and $(20,10)$, respectively. In (d), we see the diagonal lines on which the objective function $z=2x+y$ has constant values 25, 50, and 75. These lines are the *level curves* of the function $2x+y$. We note that z increases upward and to the right. In (e), we note that the level curve, $z=50$ touches the triangle ABC at the point C that has coordinates $x=20$, $y=10$. The values of z in the triangle ABC are at most equal to 50, and this value is achieved at vertex C. Hence, 50 is the minimum of z subject to the constraints (19.1).

> The relation between the nonnegative production numbers x_j and the re-source limits b_j is expressed by inequalities as follows:
>
> $$a_{i1}x_1 + a_{i2}x_2 + \cdots + a_{iN} \le b_i \ (i = 1, 2, \ldots, M)$$
>
> The total profit is equal to
>
> $$\text{Profit} = c_1 x_1 + c_2 x_2 + \cdots + c_N x_N$$
>
> The problem is to choose nonnegative production numbers x_j ($j = 1, 2, \ldots, N$) so that profit is maximum subject to the preceding inequalities.

Figure 19.4. The activity-analysis problem

the warehouses is equal to the total supply at the factories. Find a shipping schedule that meets the demands of every warehouse for minimum total cost.

To express this problem algebraically, suppose that there are m factories and n warehouses. Let a_i be the amount available at the ith factory, and let b_j be the amount required at the jth warehouse. The fact that the total supply equals the total demand is expressed by the following equation.

$$a_1 + a_2 + \cdots + a_m = b_1 + b_2 + \cdots + b_n$$

Let c_{ij} be the cost of shipping one unit of the commodity from the ith factory to the jth warehouse. We wish to find nonnegative amounts x_{ij} of the commodity to be shipped from the ith factory to the jth warehouse for minimum total cost. The fact that the supply is exhausted at every factory is expressed by the following equations.

$$
\begin{aligned}
x_{11} + x_{12} + \cdots + x_{1n} &= a_1 \\
x_{21} + x_{22} + \cdots + x_{2n} &= a_2 \\
\vdots \qquad \vdots \qquad \vdots \qquad \vdots & \\
x_{m1} + x_{m2} + \cdots + x_{mn} &= a_m
\end{aligned}
\tag{19.2}
$$

Moreover, the fact that the demand at every warehouse is met is expressed by the following equations.

$$
\begin{aligned}
x_{11} + x_{21} + \cdots + x_{m1} &= b_1 \\
x_{12} + x_{22} + \cdots + x_{m2} &= b_2 \\
\vdots \qquad \vdots \qquad \vdots \qquad \vdots & \\
x_{1n} + x_{2n} + \cdots + x_{mn} &= b_n
\end{aligned}
\tag{19.3}
$$

Furthermore, the cost to be minimized is equal to

$$
\begin{aligned}
&c_{11}x_{11} + c_{12}x_{12} + \cdots + c_{1n}x_{1n} \\
&+ c_{21}x_{21} + c_{22}x_{22} + \cdots + c_{2n}x_{2n} \\
&\quad \vdots \qquad\quad \vdots \qquad\qquad \vdots \\
&+ c_{m1}x_{m1} + c_{m2}x_{m2} + \cdots + c_{mn}x_{mn}
\end{aligned}
\tag{19.4}
$$

The transportation problem, like the diet problem and the activity-analysis problem, is a linear programming problem. In fact, the transportation problem conforms to Definition 19.2 because it asks for a minimum of the linear function (19.4) subject to the linear equalities (19.2) and (19.3).

The Simplex Method

All of the examples discussed above and, in fact, all linear programming problems can be solved by Dantzig's simplex method. In Figure 19.3, we see that the constraint set for Example 19.3 is a triangle with vertexes ABC. We can solve the problem by examining the value of the objective function at the three vertexes ABC. In a more complex example, the constraint set is an n-dimensional polytope. Again, the problem can be solved by examining the objective function at the vertexes of this polytope. Dantzig's simplex method provides a systematic algorithm for moving through a subset of the vertexes of the constraint polytope in such a way that at each step the objective function takes an improved value culminating in a vertex at which the optimum is achieved.[5]

Fifty years after its introduction, the simplex method is still the unbeaten tool for solving linear programming problems. However, the analysis of running time has raised certain questions.

Running time

It has been shown that in certain artificially contrived worst cases the running time of the simplex method is exponential.[6] However, practical experience with the simplex method in large-scale industrial applications suggests that the average running time is much faster.

The fact that the simplex method has exponential running time in the worst case led to the discovery by N. K. Karmarkar[7] in 1984 of a method for solving linear programming problems with polynomial running time. Karmarkar's method and its variants are called *interior point methods* because they obtain a sequence of points interior to the constraint polytope that leads to the optimum of the objective function.

In practice, the interior point methods have been slower than the simplex method for solving linear programming problems. However, the interior point methods have provided practical solutions to other (quadratic, etc.) mathematical programming problems.

In the next section, we discuss in more depth a specialization of the transportation problem that is called the assignment problem as well as an algorithm for its solution.

The Assignment Problem

Maximal assignment

Suppose that we are running a business that employs n workers to accomplish n different tasks. Our problem is to assign the workers to the jobs in the most productive manner. Let us assume that we have measured the daily productivity of each worker in each of the n jobs. In particular, we know that the daily benefit to the company of assigning job i to worker j is p_{ij} per day. The productivities form the following $n \times n$ array.

$$
\begin{array}{cccc}
p_{11} & p_{12} & \cdots & p_{1n} \\
p_{21} & p_{22} & \cdots & p_{2n} \\
\vdots & \vdots & \vdots & \vdots \\
p_{n1} & p_{n2} & \cdots & p_{nn}
\end{array}
$$

> Recall from Chapters 5 and 6 that a *permutation* of the numbers $1,2,...,n$ is a reordering of these numbers.

Let $j_1, j_2, \ldots, j_n$ be a permutation of the numbers $1, 2, \ldots, n$. If the ith job is assigned to worker j_i, then the total daily productivity is

$$
p_{1j_1} + p_{2j_2} + \cdots + p_{nj_n}
$$

We want to find the assignment of workers to jobs that *maximizes* the total productivity.

Minimal assignment

An alternate way of formulating the assignment problem is to consider the daily *cost* of employing each worker to do the various jobs. Let c_{ij} be the cost entailed in assigning the jth worker to accomplish the ith task. The costs form the $n \times n$ array

$$
\begin{array}{cccc}
c_{11} & c_{12} & \cdots & c_{1n} \\
c_{21} & c_{22} & \cdots & c_{2n} \\
\vdots & \vdots & \vdots & \vdots \\
c_{n1} & c_{n2} & \cdots & c_{nn}
\end{array}
\tag{19.5}
$$

Consider the special assignment in which the ith job is assigned to the ith worker. In this case, the total cost C_0 to carry out all the tasks is

$$
C_0 = c_{11} + c_{22} + \cdots + c_{nn}
\tag{19.6}
$$

In general, if the ith job is assigned to worker j_i, then the total cost C is

$$
C = c_{1j_1} + c_{2j_2} + \cdots + c_{nj_n}
\tag{19.7}
$$

We want to find the assignment of workers to jobs that *minimizes* the total cost.

For technical reasons, we will assume that the numbers p_{ij} or c_{ij} are integers. This is not a significant restriction because we can take units of productivity or cost as small as necessary.

One way to solve these problems is to evaluate every possible assignment of the workers to the jobs. The problem with this solution is that for moderate sized values of n, the number of assignments to evaluate is beyond the capabilities of even the fastest computers. For example, if n is equal to 50, then the number of assignments, 50!, is approximately equal to 3×10^{64}. At the rate of one million evaluations per second, this job would require about 10^{51} years. For comparison, the age of the universe is said to be approximately 10^{10} years. We will discuss an algorithm that reduces the time for this task to less than one second.

Notice that an assignment problem is a special kind of linear programming problem. To see this, we must modify the problem slightly so that *job sharing* is allowed. That is, we will permit an employee to divide his time between several tasks. Specifically, let x_{ij} be nonnegative numbers representing the fraction of time that the jth worker spends on the ith job. Since every worker works full time, we must have

$$x_{1j} + x_{2j} + \cdots + x_{nj} = 1$$

for i between 1 and n; and since every job is a full-time job, we must have

$$x_{i1} + x_{i2} + \cdots + x_{in} = 1$$

for j between 1 and n. The minimal assignment problem can be interpreted as a transportation problem in which every factory has total supply equal to 1 and every warehouse has demand equal to 1. The maximal assignment problem differs from a transportation problem in that the maximal problem seeks a maximum of the objective function whereas a transportation problem seeks a minimum, but we will see that this difference is superficial. The objective function for the maximal assignment problem is

$$
\begin{aligned}
& p_{11}x_{11} + p_{12}x_{12} + \cdots + p_{1n}x_{1n} \\
&+ p_{21}x_{21} + p_{22}x_{22} + \cdots + p_{2n}x_{2n} \\
&\qquad \vdots \qquad\qquad \vdots \qquad\qquad \vdots \\
&+ p_{n1}x_{n1} + p_{n2}x_{n2} + \cdots + p_{nn}x_{nn}
\end{aligned}
\tag{19.8}
$$

The assignment problem is a special linear programming problem. The simplex method, which is an all-purpose solution method for linear programming problems, is available for the solution. However, we prefer to use a special purpose algorithm, called the *Hungarian method*, for the solution of the assignment problem. This method was devised by Harold Kuhn,[8] who called it the Hungarian method because it is based on the work of two Hungarian mathematicians, Dénes Kőnig and E. Egerváry. In the next section, we discuss a Kőnig theorem that is used in the Hungarian method.

Kőnig's theorem

Suppose that a rectangular array—we will call it a 0-array—consists of zeros and blanks. Here is an example of such a 0-array.

0		0		0
		0		
	0	0		
	0	0	0	
		0		

Definition 19.4. A subset of the zeros in a 0-array is called *independent* if no two zeros in the subset belong to the same row or column.

In our example, the subset consisting of the zeros marked * is independent.

0*		0		0
		0*		
	0*	0		
	0	0	0*	
		0		

In this example, we see that there is a subset of four independent zeros. Moreover, by testing all possibilities one can see that there do not exist five independent zeros. A set of independent zeros with the largest possible number of elements is called a *maximal* set of independent zeros.

In a 0-array, by a *line* we mean either a row or a column

Definition 19.5. A set of lines that includes all the zeros in a 0-array is called a *cover* of the 0-array.

In our example, the set of all the rows, or the set of all of the columns, is an example of a cover consisting of five lines. However, in this example, it is possible to find a four-line cover consisting of one row and three columns as indicated by the arrows $\Leftarrow$ and $\Uparrow$.

0		0		0	$\Leftarrow$
		0			
	0	0			
	0	0	0		
		0			
	$\Uparrow$	$\Uparrow$	$\Uparrow$		

By examining all possibilities, we discover that, in this example, there does not exist a cover with fewer than four lines. A cover with the smallest possible number of lines is called a *minimal* cover. Now we are ready to state Kőnig's theorem.[9]

Theorem 19.6 (Kőnig's theorem). *For every 0-array, the number of lines in a minimal cover is the same as the maximal number of independent zeros.*

We will not discuss the proof of this theorem. We have verified that it is correct for our example. The reader may wish to devise other examples.

Reduction

We wish to convert a given assignment problem into an equivalent assignment problem that can be solved immediately. This *easiest* $n \times n$ assignment problem has the following features.

1. The easiest problem is a *minimal* assignment problem—the problem of minimizing total cost.

2. In the easiest problem, the c_{ij} are all nonnegative.

3. If we replace the nonzero numbers in the n by n array (19.5) by blanks, then what remains is a 0-array. In the easiest case, this 0-array contains an independent set of n zeros.

The reason that these three properties make the problem easy is the following.

1. The total (19.7) to be minimized is nonnegative because in 2 above we assume that all the numbers c_{ij} are nonnegative.

2. The n independent set of zeros in 3 above corresponds to an assignment because there is exactly one of these zeros in every row and every column of the n by n array of the numbers c_{ij}. For this assignment, the total (19.7) is zero—in fact, every term in the sum is zero. But since all terms in (19.7) are ↩ Dangerous curve! nonnegative, *no assignment can give a total that is less than zero.* Thus, we have found the optimal assignment, and the corresponding total (19.7) is equal to zero.

To reduce an assignment problem to the easiest case, we follow certain steps that we will illustrate with the following example.

Example 19.7. *Optimal assignment problem.* Cal has been operating a one-man drapery business for several years but now wants to continue the business with the help of his friends Ada, Ben, Dot, and Eli. They have decided that there are five positions needed to run the business: an Adman, a Buyer, a Cutter, a Dyer, and an Edger. The five have determined their productivities (Figure 19.5(a)) if employed in each of the five positions. Determine an assignment of the five friends to the five positions in the manner that maximizes the sum of the productivities.

We will do much more than just solve this special problem. In fact, we will describe a method that can be used to solve any assignment problem whatever. The idea of this method is to reduce the given assignment problem to the *easiest* assignment problem as described above—a minimal assignment such that (1) all costs are nonnegative and (2) a particular assignment has zero total cost.

	Ada	Ben	Cal	Dot	Eli	max
Ad	7	2	8	0	7	8
Buy	2	3	8	1	3	8
Cut	4	6	8	3	5	8
Dye	2	6	8	7	5	8
Edge	1	0	4	1	0	4

(a) Maximum problem

	Ada	Ben	Cal	Dot	Eli
	1	6	0	8	1
	6	5	0	7	5
	4	2	0	5	3
	6	2	0	1	3
	3	4	0	3	4
min	1	2	0	1	1

(b) Minimum version

	Ada	Ben	Cal	Dot	Eli	
Ad	0	4	0	7	0	⇐ 0
Buy	5	3	0	6	4	−2
Cut	3	0	0	4	2	−2
Dye	5	0	0	0	2	−2
Edge	2	2	0	2	3	−2
		⇑	⇑	⇑		
	0	+2	+2	+2	0	

(c) Cover found

	Ada	Ben	Cal	Dot	Eli
Ad	0	6	2	9	0*
Buy	3	3	0*	6	2
Cut	1	0*	0	4	0
Dye	3	0	0	0*	0
Edge	0*	2	0	2	1

(d) Optimal assignment

Figure 19.5. Data and calculations for Example 19.7.
Table (a) shows the daily productivity of each worker at each job. The problem is to assign workers to jobs so that the total productivity of all workers is maximum. The numbers in the right border of (a) are the maximums of the corresponding rows.

In **Table (b)**, we replace the original maximum problem with an equivalent problem of finding the assignment that *minimizes* a sum of nonnegative quantities. Table (b) is obtained from (a) by a two-step process. (1) We replace every number in (a) by its negative. This is done to convert the original maximum problem into a minimum problem. (2) To the resulting negative elements we add the appropriate row-maximum from (a). Adding the same number k to each element of a row does not change the optimal assignment because for all assignments the sum to be minimized is increased by the same number k. The numbers in the bottom border of (b) are the minimums of the corresponding columns.

Table (c) is obtained by subtracting the column minimums of (b) (shown in the bottom border) from each element of the corresponding column of (b). The arrows in (c) show a cover of four lines consisting of one row and three columns.

Table (d) is obtained from (c) in two steps. (1) The numbers in the left border of (c) are subtracted from elements in the corresponding rows. (2) The numbers in the bottom border of (c) are added to the elements in the corresponding columns.

In (d), the zeros marked with * are an independent set of five zeros. The elements 0^* in (d) mark the desired optimal assignment, giving a solution to Example 19.7.

1. If the given problem is a maximal assignment problem, convert it to a minimal problem with $c_{ij} = -p_{ij}$. The optimal assignment for the new problem is clearly the same as for the old problem.

2. If we add (or subtract) a constant amount k from each element in a row (or column) of the cost array (19.5), then the optimal assignment is unchanged because the cost of *every* assignment is changed by the amount k. In fact, total cost of a particular assignment includes exactly one element from the altered row (or column) of the cost array (19.5); therefore, the total cost of every assignment is increased (or decreased) by the amount k.

   Below in step 3, we repeat this process (adding or subtracting from various rows and columns) until the given assignment problem is converted to a problem of the *easiest* type—discussed on page 287, but first we use the same technique in (a) and (b) to reduce the given minimal problem to an equivalent minimal problem in which (1) all elements of the cost array are nonnegative and (2) the cost array contains at least one zero in every row and every column.

   (a) For each row, subtract the minimum of the row from each element in the row.

   Figure 19.5(b) shows the result of applying steps 1 and 2(a) to the data from Example 19.7 in Figure 19.5(a).

   (b) For each column, subtract the minimum of the column from each element of the column.

   Figure 19.5(c) shows the result of applying step 2(b) to Figure 19.5(b).

   After carrying out steps (a) and (b), note that there is at least one zero in every row and every column of the cost array.

3. Let (19.5) denote the array as modified (if necessary) in steps 1 and 2. Ignoring the nonzero elements, consider the array (19.5) as a 0-array. According to König's theorem, exactly one the the following alternatives holds.

   - There exists a set of m $(m < n)$ covering lines.
   - There exists a set of n independent zeros.

   Depending on which of these alternatives holds, we proceed to step 4 or 5.

4. If there exists a set of m $(m < n)$ covering lines, proceed as follows:

   (a) Find the smallest positive number s that is in the array (19.5) but *not* covered by the lines—not in the covered rows or columns.

   (b) Subtract the number s from every element in the **rows** that *are not* in the cover.

   (c) Add the number s to every element in the **columns** that *are* in the cover.

In Figure 19.5(c), the three arrows in the bottom border and the arrow in the right border mark a cover with four lines. The minimum entry outside of the cover is 2. In the right border there is a -2 for every row not in the cover, which indicates that 2 must be subtracted from each of those rows. In the bottom border there is a $+2$ for every column in the cover, which indicates that 2 must be added to every element in those columns. The result of this computation is shown in Figure 19.5(d).

Let us examine the effect of this procedure on the cost of a particular assignment, the *test assignment*, in which the ith job is done by the ith worker. The cost C_0 of this assignment is shown in (19.6).

In general, suppose that the number of rows in the cover is r and the number of columns is c. We subtracted s from the items in $n - r$ rows and we added s to c columns. In doing so we altered the total cost C_0 of the test assignment, shown in (19.7), by the amount

$$s(c - (n - r)) = s(c + r - n) = s(m - n) < 0$$

Since this amount is negative, it means that we have reduced the total cost C_0 of the test assignment—indeed, of any assignment. Since we assumed that the input numbers (p_{ij} or c_{ij}) are integers, and since we have performed only addition and subtraction of these numbers, the number s must be a positive integer. At each application of this procedure, we reduce C_0 by a positive integer.

Remember that c, r, and m are, respectively, the number of columns, the number of rows, and total number of lines in the cover; hence, we have $c + r = m$.

After this reduction, we **return to step 3** followed by either step 4 or 5. However, notice that we can return to step 4 only finitely many times because, on the one hand, each repetition of step 4 causes C_0 to decrease by at least 1, and, on the other hand, C_0 must be nonnegative. Note that if C_0 becomes zero, then the test assignment—which assigns the ith job to the ith worker—is an optimal assignment.

The test assignment is discussed in order to establish that the process terminates. In practical computations, we do not need to compute the cost of the test assignment.

5. If there exists a set of n independent zeros, then *we are finished*. As discussed above, we have found an assignment for which the total (19.7) is zero. Since for every assignment the total (19.7) is nonnegative, and since we are solving a minimum problem, this assignment is optimal because its total cost is zero.

In Figure 19.5(d), the zeros marked * are a set of five independent zeros. Thus, a solution to Example 19.7 is:

Ada is the Edger,	1 unit
Ben is the Cutter,	6 units
Cal is the Buyer,	8 units
Dot is the Dyer,	7 units
Eli is the Adman,	7 units
	Total	29 units.

In the solution of Example 19.7 in Figure 19.5, we apply step 3 twice. The first time we proceed to step 4, and the second time we find the optimal assignment at step 5. In general, we must return to step 3 multiple times. Each time we perform step 4 we must return to step 3, which tells us whether to repeat step 4 or to terminate the process by going to step 5. Eventually, we must arrive at an optimal assignment via step 5.

Note that although Kőnig's theorem assures us that it is possible to do so, we have not defined an algorithm for finding a cover with fewer than n lines in 4, or for finding a set of n independent zeros in 5. In carrying out these computations for small problems (up to about 20×20), it is feasible to find the required covering lines or independent zeros by trial-and-error.

We have had a glimpse of mathematical programming, a very large field.[10] In the next chapter, we will see similar techniques applied to games of business, war, and recreation.

Chapter 20

The Play's the Thing

Confucius said: "What can be done with a man who stuffs his face with food all day, without exercising his mind. He could at least play cards or chess or something. It would be better than nothing."

—CONFUCIUS, *Analects*[1] 17:22

There are games for recreation and gambling, but there are also games of business and warfare. There are even games that exemplify perplexing ethical questions.

The mathematical theory of games first achieved recognition in 1944 with the publication of *The Theory of Games and Economic Behavior* by von Neumann and Morgenstern.[2] This work seeks a mathematical understanding of competitive situations.

General Concepts

The central concepts of the theory of games are *payoff* and *strategy*.

Strategy

In any game, the players are given choices of actions to take. Sometimes these choices are very complex. In actual cases, these choices might be made over a considerable time period, and frequently choices are not made until a player sees choices made by the other player(s). For example, in chess the players move by turns in ways that may not have been planned at the start of the game. Nevertheless, a *strategy* of a player is defined as a list of what that player will do in every possible contingency of the game. A strategy in this sense is also called a *pure strategy* in contrast to a *mixed strategy*, which will be defined shortly. We will consider games in which the number of possible strategies is finite, but for many games this number is extremely large. For example, in the case of chess, a strategy is a purely theoretical object because it isn't feasible to list every contingency of a game of chess. At the opposite extreme of complexity, "Bet two dollars on Blue Lightning in the third race" is a strategy. If a player makes his strategy known to an assistant, then that person can function as the player's proxy without further consultation with the player.

Payoff

Depending only on the strategies chosen by the players, each player receives a *payoff*. The payoff can be any sort of quantifiable benefit, but we will refer to the payoff as a sum of money. When a player is assigned a negative payoff, it means that he pays money instead of receiving it. For example, in chess the payoffs are $+1$ for a win, -1 for a loss, and 0 for a draw. Both players have complete knowledge of the payoffs that are associated with all possible outcomes of the game. Before selecting a strategy, a player could, at least in theory, examine the payoffs associated with all possible strategies—her own strategies in combination with the other players' strategies. However, in chess, for example, there are so many strategies that this is not a feasible procedure. For some games, known as *zero-sum games*, the sum of the payoffs to all the players is zero—the winnings of some players are balanced by other players' losses. Gambling games are generally zero-sum games. In other games the sum of payoffs may be positive. Business games are of this sort when some economic good is created during the game. War causes destruction, and therefore it is a game in which the sum of the payoffs is negative.

Cooperative and noncooperative games

Games can be *cooperative* or *noncooperative*. In a cooperative game, the players are allowed to form coalitions with other players. Coalitions can make binding agreements concerning their strategies. Furthermore, they can agree to side payments among themselves. Cooperative games are a mathematical model of business and warfare. In noncooperative games no coalitions are permitted. Two-person zero-sum games are always noncooperative; they model mainly parlor games and gambling.

The game of Odd Man Out

When we speak of *n*-person games, we mean games with n players where n is greater than two. We will discuss one example of this type of game—a three-person game. This example comes from the theory of *cooperative* games.

Example 20.1 (Odd Man Out). Ada, Ben, and Cal play the following game. Each one of the three must vote for one of the other two. Any two of them who vote for each other collect $100 each from the third. Before the voting, there is a period of time in which to form coalitions and agree on side payments.

If we were present during the negotiation period, we might hear a conversation something like this.

Ada: Ben, let's vote for each other and each collect $100 from Cal.

Ben: I agree, Ada.

Cal: Ben, I'll give you a better deal. Let's vote for each other, and each collect $100 from Ada. I'll give you a side payment of $50 from my winnings.

Ben: I agree, Cal.

Ada: Cal, if you vote for me, you won't have to make a side payment. You can keep everything you get from Ben.

Cal: I agree, Ada.

Ben: Cal, if you vote for me, I'll give you a side payment of $50.

.   .   .   .

Probably the voices become more excited as the negotiation continues, but to no avail. At some point they are all talking at once. We see that the coalitions are unstable, and that there is no rational behavior that has any conclusive benefit. Even if one of the three is an expert in cooperative *n*-person games, he or she has no advantage over the others. Ada, Ben, and Cal have put themselves in a very unpleasant and confusing situation.

Odd Man Out is a metaphor for the failure of human fellowship; it shows that rational self-interest has a dark side.

As we move from the theory of *n*-person games to two-person non-zero-sum games and finally to two-person zero-sum games, the mathematical modeling of economics and warfare becomes less accurate; on the other hand, the theory becomes more of a guide for the rational behavior of the individual players. There is an extensive theory of *n*-person games, but for the rest of this chapter we will consider two-person games exclusively.

Two-person Games

The rules of a two-person game can be very complex, but despite their seeming variety, most of these games can be reduced to a particular type of game called a *matrix game*. Central to the game is a *rectangular payoff matrix*.

matrix. A rectangular array of numbers.

We use a game called the *Prisoners' Dilemma* to introduce the concept of a payoff matrix. Like *Odd Man Out*, this game also has ethical interpretations. Ada and Ben have been detained by the state police of Lemuria because they own transistor radios, a crime in Lemuria punishable by one year in prison. However, the Lemurians believe that Ada and Ben are guilty of spying, a much more serious crime. Each of them is interviewed separately by the police, and each is offered two alternatives:

1. Confess and implicate your partner.

2. Deny all charges.

They are informed that, depending on their choices, their punishments will be as follows. If Ada confesses to spying and implicates Ben and he denies the charge, then she goes free and he serves five years in prison. Similarly, if Ben confesses and implicates Ada and she denies the charge, then he goes free, and she serves five years in prison. If both confess, then both serve three years in prison. If both deny, then both serve one year in prison for possessing the transistor radios. Ada's and Ben's choices are shown in Figure 20.1.

BEN

		confess	deny
ADA	confess	$(-3, -3)$	$(0, -5)$
	deny	$(-5, 0)$	$(-1, -1)$

Figure 20.1. The Prisoners' Dilemma. Ada chooses a row (confess or deny) and Ben chooses a column (confess or deny). Each makes his or her choice without knowing the other's choice. For each item of the payoff matrix, the first number is the prison sentence of Ada, and the second number is the sentence of Ben. For example, if Ada confesses and Ben denies, then Ada goes free and Ben spends five years in prison. The payoffs are negative because by convention a positive payoff indicates desirable units of utility—often a sum of money.

Games like the Prisoners' Dilemma occur both in literature and ordinary life. A famous instance occurs in Puccini's opera, *Tosca*. The game is played between Tosca, a beautiful Roman singer, and Scarpia, the evil chief of police. Tosca and Scarpia have made a deal: She agrees to sleep with him if he will release from prison her lover, Cavaradossi, who is awaiting execution. Tosca and Scarpia must each decide whether to honor this agreement. In fact, they decide to betray each other. Tosca decides to stab Scarpia to death rather than submit, but it is too late because Scarpia has already given the order to carry out the execution of Cavaradossi.

The Prisoners' Dilemma illustrates the concept of *equilibrium point*. In Figure 20.1, the point $(-3, -3)$, the payoff in case both Ada and Ben confess, is an equilibrium point because if Ada confesses, then Ben can do no better than to confess, and, similarly, if Ben confesses, then Ada can do no better than to confess. The paradox of the Prisoners' Dilemma is that this equilibrium point—which can be characterized as the *double double cross*—is an undesirable outcome. In contrast, the point $(-1, -1)$, that represents the more desirable outcome in which Ada and Ben both deny, is not an equilibrium point because if Ada denies, then Ben can improve his payoff by confessing and implicating Ada, and, similarly, if Ben denies, then Ada can improve her payoff by confessing.

The equilibrium point shows an interesting structure of the game, but it is not very helpful to players who must decide whether to confess or deny. Nevertheless, it would be interesting, at least from a theoretical point of view, if every similar game had an equilibrium point. This will turn out to be true if some further definitions are made, but as matters stand now the following example, a two-person zero-sum game, shows that there need not be an equilibrium point.

Example 20.2 (coin matching game). Instead of using coins, Ada and Ben write either *heads* or *tails* on slips of paper. They compare the slips. If they match, then Ben must pay Ada one dollar; otherwise, Ada must pay Ben one dollar.

The payoff matrix for this game is shown in Figure 20.2. It is evident that there is no equilibrium point because the losing player always finds it advantageous to switch. However, the assertion that there always exists an equilibrium point can be restored by defining *mixed strategy*. Before we define mixed strategy in general, we illustrate it for the coin matching game. Ada and Ben use mixed strategies if, for example, each one tosses a coin (out of the sight of the other player) and then each writes on his or her slip of paper the result of the toss. If we assume a fair toss of an unbiased coin, then the effect of this coin-tossing procedure is that each player plays *heads* or *tails*, each with probability $1/2$.[3] Since the random strategy choices of Ada and Ben are independent, we find that the probability, for example, that both Ada and Ben choose heads is equal to $1/2 \cdot 1/2 = 1/4$. The expected payoff—that is, the expected amount that Ben pays Ada—is computed by multiplying each payoff by the probability that it will occur.[4]

$$\text{Expected payoff} = \frac{1}{2} \cdot \frac{1}{2} \cdot 1 + \frac{1}{2} \cdot \frac{1}{2} \cdot (-1) + \frac{1}{2} \cdot \frac{1}{2} \cdot (-1) + \frac{1}{2} \cdot \frac{1}{2} \cdot 1 = 0$$

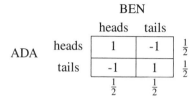

Figure 20.2. The coin matching game. Because this is a zero-sum game, the amount received by one player is equal to the amount paid by the other player; therefore, we do not need to specify the payoffs for each of the two players. The upper left entry 1 means that if Ada and Ben select *heads* then Ben pays Ada one dollar. The $1/2$'s in the right and bottom borders refer to the *mixed strategy* in which Ada and Ben each play heads or tails with probabilities $1/2$ by tossing a coin.

Definition 20.3. We say that a player uses a mixed strategy if she plays her various pure strategies according to certain probabilities. (The sum of these probabilities must equal 1.)

The word *mixed* does not imply that more than one strategy has a nonzero probability. If a player decides with certainty to play one particular strategy, then that is the mixed strategy in which one strategy receives probability one and the probabilities of all other strategies are zero. The implementation of a mixed strategy involves consulting a chance device—tossing a coin, throwing dice, or consulting a table of random numbers. The latter alternative has the advantage that it is easier to fine tune the probabilities.

The mixed strategy for the coin matching game discussed above is an equilibrium point because neither player can achieve an advantage by departing from this mixed strategy. An equilibrium point requires that no advantage can be gained by switching from the equilibrium point, assuming that the other player remains at the equilibrium point, and we see that this is true. Indeed, in this game the equilibrium point is also a *minimax* point, a stronger concept we will define in the next section.

It is true that for any noncooperative (not necessarily zero-sum) game, there is an equilibrium point in the sense that there are mixed strategies such that neither player can increase his expected payoff by departing from his equilibrium strategy *provided that his opponent does not depart from the equilibrium strategy.* We see from the example of the Prisoners' Dilemma that knowing the equilibrium point (in which both players confess) does not lead to the more desirable outcome (in which both players deny). The existence of equilibrium points for this class of games was discovered by John Nash (1928–) in 1951. He was awarded the 1994 Nobel Prize in economics for this and other contributions to the theory of games.

> *Q. Why didn't Nash get the Nobel Prize in mathematics?*
>
> *A. There is no Nobel Prize in mathematics. There are speculations as to why this is so. The most outrageous rumor is that a mathematician ran off with Alfred Nobel's wife, but this is without foundation because Nobel was a bachelor.*

Two-person Zero-sum Games

In a matrix game, the row player, also known as the first player, selects a row, and the column player, also known as the second player, selects a column. In a two-person zero-sum game, the entry of the payoff matrix corresponding to the chosen row and column is the amount that the first player receives from the second player. A negative amount represents money paid by the first player to the second player.

Definition 20.4. A *saddle point* of a two-person zero-sum payoff matrix is an entry that is the minimum of its row and the maximum of its column.

> *Q. Why is it called a saddle point?*
>
> *A. The primary mathematical usage of the term* saddle point *is in geometry. On an actual saddle an imaginary path from stirrup to stirrup achieves a maximum elevation at the same point (the saddle point) at which a path from pommel to cantle achieves a minimum.*

Not all matrix games have saddle points. In Figure 20.3, the 1 in the upper left corner of the payoff matrix is a saddle point. The row player can do no better than to

1	2
0	3

Figure 20.3. The upper left entry of this payoff matrix is a saddle point.

select the first row because by doing so he is assured of receiving at least one dollar. On the other hand, the column player can do no better than to select the first column, because by doing so he is assured that he will not have to pay any more than one dollar. Both players have good reason to play the saddle point if they believe that their opponent is acting with rational self-interest.

Definition 20.5. For a two-person zero-sum game, if there is a number v (positive or negative) such that

1. There is a strategy (pure or mixed) for the first player that assures that her (expected) payoff is no less than v,
   and
2. There is a strategy for the second player that assures that his (expected) payoff is no less than $-v$,

then we say that v is the *value* of the game, and that the strategies mentioned above are *optimal strategies*. (In the case of *mixed* optimal strategies, the payoffs mentioned in 1 and 2 are *expected* payoffs.)

The value of the game in Figure 20.3 is one dollar, and the optimal strategies are:

1. The first player chooses the first row.

2. The second player chooses the first column.

A game with a saddle point is not an interesting pastime when the optimal strategies are known. As we will see, tic-tac-toe and chess belong to a class of games that have saddle points. One can learn optimal strategies for tic-tac-toe but not for chess because the number of strategies for chess is inconceivably large. On the other hand, games with mixed optimal strategies include some popular gambling games.

The traditional Italian game of Morra is one such game. In Morra, the two players each hold up between zero and four fingers and simultaneously shout at each other a number between 0 and 8. A player wins the amount that he calls if he calls the total number of fingers shown by both players. It is a draw if neither player guesses correctly. According to an old Italian proverb, the outer limit of trust is to play Morra in the dark for money. We will discuss a simplified version of Morra in which the players each show 1, 2, or 3 fingers.[5] The payoff matrix and optimal strategies for three-finger Morra are shown in Figure 20.4.

It is straightforward to check that the strategies shown in Figure 20.4 are optimal. To do so, show that if Ben uses the mixed strategy that is claimed to be optimal, then no matter which pure strategy Ada uses, her expected payoff is at most equal to zero. By symmetry, it will also be true that if Ada uses the claimed optimal strategy, then Ben's expected loss from each pure strategy is not less than zero. The following is the computation of Ada's expected payoff for each of her pure strategies, that is, for each of her row choices.

The computation in Figure 20.5 shows that none of Ada's pure strategies gives her a payoff greater than zero, provided that Ben uses the claimed optimal strategy.

We can carry out a similar computation with the roles of Ada and Ben reversed, but the following argument shows that the calculation shown in Figure 20.5 is already sufficient. Indeed, the payoff matrix in Figure 20.4 has the property that if rows and columns are interchanged, then every entry is replaced by its negative. This means that if Ada plays her claimed optimal strategy, then the expected payoff for each of Ben's pure strategies is as shown in Figure 20.5 with plus and minus signs interchanged. This implies that Ben cannot achieve a payoff lower than zero with

BEN

	12	13	14	23	24	25	34	35	36	
12	0	2	2	−3	0	0	−4	0	0	0
13	−2	0	0	0	3	3	−4	0	0	0
14	−2	0	0	−3	0	0	0	4	4	$\frac{5}{12}$
23	3	0	3	0	−4	0	0	−5	0	0
24	0	−3	0	4	0	4	0	−5	0	$\frac{4}{12}$
25	0	−3	0	0	−4	0	5	0	5	0
34	4	4	0	0	0	−5	0	0	−6	$\frac{3}{12}$
35	0	0	−4	5	5	0	0	0	−6	0
13	0	0	−4	0	0	−5	6	6	0	0
	0	0	$\frac{5}{12}$	0	$\frac{4}{12}$	0	$\frac{3}{12}$	0	0	

ADA (left border label)

Figure 20.4. Ada and Ben play *three-finger Morra*. Pure strategies are shown outside the box in the left and upper borders of the matrix. The first digit is the number of fingers shown, and the second digit is the number called. The right and bottom borders show optimal mixed strategies: *Always call 4. Show 1, 2, and 3 fingers with probabilities* 5/12, 4/12, and 3/12, *respectively.* The value of the game is zero. Ada and Ben found the game boring because when players use these optimal strategies, no money ever changes hands.

$$2 \cdot \tfrac{5}{12} + 0 \cdot \tfrac{4}{12} - 4 \cdot \tfrac{3}{12} = -\tfrac{2}{12}$$
$$0 \cdot \tfrac{5}{12} + 3 \cdot \tfrac{4}{12} - 4 \cdot \tfrac{3}{12} = 0$$
$$0 \cdot \tfrac{5}{12} + 0 \cdot \tfrac{4}{12} + 0 \cdot \tfrac{3}{12} = 0$$
$$3 \cdot \tfrac{5}{12} - 4 \cdot \tfrac{4}{12} + 0 \cdot \tfrac{3}{12} = -\tfrac{1}{12}$$
$$0 \cdot \tfrac{5}{12} + 0 \cdot \tfrac{4}{12} + 0 \cdot \tfrac{3}{12} = 0$$
$$0 \cdot \tfrac{5}{12} - 4 \cdot \tfrac{4}{12} + 5 \cdot \tfrac{3}{12} = -\tfrac{1}{12}$$
$$0 \cdot \tfrac{5}{12} + 0 \cdot \tfrac{4}{12} + 0 \cdot \tfrac{3}{12} = 0$$
$$-4 \cdot \tfrac{5}{12} + 5 \cdot \tfrac{4}{12} + 0 \cdot \tfrac{3}{12} = 0$$
$$-4 \cdot \tfrac{5}{12} + 0 \cdot \tfrac{4}{12} + 6 \cdot \tfrac{3}{12} = -\tfrac{2}{12}$$

Figure 20.5. Verification of an optimal mixed strategy for the game of *three-finger Morra*. In each product above, the left factor is a payoff from Figure 20.4, and the right factor is one of the three probabilities 5/12, 4/12, 3/12.

any of his pure strategies. (Remember that the payoff represents an amount that Ben must pay Ada—Ben is the *minimizing* player.)

Since, using the claimed optimal strategies, Ada is assured of an expected payoff of at least zero and Ben is assured that the expected payoff will not exceed zero, we conclude that these strategies are indeed optimal and that zero is the value of the game.

> *Q. In an actual game of Morra, the play goes too fast for one to do any calculations. How could the knowledge of the optimal strategies give practical assistance in playing the game?*
>
> *A. One practical solution is to memorize in advance a list of plays constructed by consulting a chance device. For example, suppose that we have a table of random numbers between 0 and 1. We divide this interval into disjoint subintervals of length 5/12, 4/12, and 3/12, respectively. We choose numbers sequentially from the random number table. If x occurs in this sequence of random numbers, then a play in our list is constructed from x as follows.*
>
> - *If x satisfies $0 \leq x < \frac{5}{12}$, show 1 finger, call 4.*
> - *If x satisfies $\frac{5}{12} \leq x < \frac{5}{12} + \frac{4}{12} = \frac{9}{12}$, show 2 fingers, call 4.*
> - *If x satisfies $\frac{9}{12} \leq x \leq \frac{9}{12} + \frac{3}{12} = 1$, show 3 fingers, call 4.*
>
> *We must memorize a list of plays long enough to cover the entire session of play.*

The minimax theorem

At this point it is natural to ask whether every zero-sum two-person game has a value and optimal strategies. The affirmative answer to this question was discovered by von Neumann in 1928.[6] This result is called the minimax theorem.

Theorem 20.6 (minimax theorem). *Every two-person zero-sum game has a value and optimal strategies for both players.*

This theorem provides guidance for players of two-person zero-sum games. Since it ensures optimal strategies exist, the rational way to play is to find and implement these strategies. This is in sharp contrast to *n*-person and non-zero sum games such as Odd Man Out and the Prisoners' Dilemma in which rational self-interest leads to unresolvable difficulties. We will prove the minimax theorem in the special case in which the payoff matrix has two rows and three columns; however, the methods that we use can used be to prove the general case.

In Figure 20.6, the payoff matrix for a 3×2 game is written using double subscripts; the first subscript is the row number and the second subscript is the column number.

BEN

m_{11}	m_{12}	m_{13}
m_{21}	m_{22}	m_{23}

ADA

Figure 20.6. Payoff matrix for a 2×3 game. Ada chooses one of the two rows, and, independently, Ben chooses one of the three columns. The payoff matrix shows the amount that Ben pays Ada depending on the row and column selected. A negative amount means that Ada pays Ben.

Center of mass

We begin by looking at the game from Ben's point of view. In Figure 20.7, we represent the columns (Ben's pure strategies) as points in two-dimensional space.

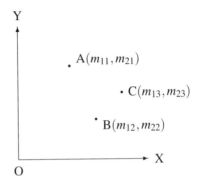

Figure 20.7. Ben's pure strategies.

Now let us represent Ben's mixed strategies in the same graphical manner in Figure 20.8. As an example, suppose that he plays the three rows with equal probabilities—each row with probability $1/3$. This strategy is represented by the point M with the coordinate representation

$$\left(\frac{m_{11}}{3} + \frac{m_{12}}{3} + \frac{m_{13}}{3}, \frac{m_{21}}{3} + \frac{m_{22}}{3} + \frac{m_{23}}{3} \right)$$

The point M is the *center of mass* of a system of three equal masses placed at points A, B, and C. This generalizes to two dimensions the concept of one-dimensional center of mass.[7]

Fact 20.7. *The center of mass of three arbitrary masses (not all zero) placed at points A, B, and C, respectively, always is located in the triangle ABC. Moreover, any point in the triangle, including the boundary of the triangle, is the center of mass of suitable masses, with total mass equal to 1, placed at points A, B, and C.*

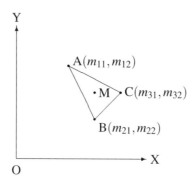

Figure 20.8. The totality of Ben's mixed strategies are represented by points that fill out the entire triangle ABC, and for each such point there is precisely one mixed strategy. Similarly, the center of mass of a system of three mass points at A, B, and C can be any point in the triangle ABC. The point M represents Ben's mixed strategy in which he plays each column with probability $1/3$; on the other hand, M is the center of mass of three equal masses located at A, B, and C.

Ben's mixed strategy with probabilities q_1, q_2, and q_3 ($q_1 + q_2 + q_3 = 1$) is represented in Figure 20.8 by the same point as the center of mass of a system of three masses q_1, q_2, and q_3 placed at points A, B, and C.

By suitably choosing his probabilities, q_1, q_2, and q_3, Ben can place his mixed strategy anywhere in the triangle ABC. He makes his choice based on a pessimistic assumption. He assumes that Ada will make the best choice available to her. Since Ada's choice is to pick either the x-component or the y-component of the point that represents Ben's mixed strategy, the pessimistic assumption is that she will pick the larger component—remember that Ada seeks the maximum payoff and Ben seeks the minimum. For each point in the triangle ABC, Ben looks at the pessimism function, the maximum of x and y, which we denote $\max(x, y)$. He chooses his mixed strategy, q_1, q_2, and q_3, to place the point representing his mixed strategy at a point in triangle ABC such that $\max(x, y)$ is as small as possible. Let v denote this minimum value of $\max(x, y)$ over the triangle ABC. This choice is Ben's optimal mixed strategy; it assures him that the expected payoff will be no more than v.

Now we switch to Ada's point of view. She knows that Ben can prevent the expected payoff from exceeding v, but she needs to make sure that her expected payoff is at least this large. To proceed with Ada's analysis, we need to know certain facts from analytic geometry.

Fact 20.8. *Let p_1 and p_2 be nonnegative numbers such that $p_1 + p_2 = 1$, and let v be arbitrary. The points with coordinates (x, y) such that $p_1 x + p_2 y = v$ lie on a straight line l containing the point (v, v).*

- *Points above or to the right of l satisfy $p_1 x + p_2 y \geq v$.*

- *Points below or to the left of l satisfy $p_1 x + p_2 y \leq v$.*

Figure 20.9 shows Ada can find an optimal mixed strategy by selecting a certain line l. The caption of Figure 20.9 shows the details of this construction, explains why v is the value of the game, and concludes the proof of Theorem 20.6 in this special case. The argument used here can be modified to prove Theorem 20.6 in general.

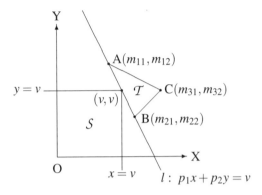

Figure 20.9. The region $\mathcal{T}$ is the triangle with vertexes A, B, and C; $\mathcal{T}$ represents the totality of Ben's mixed strategies. The region S is the inverted quadrant bounded by the half lines $x = v$ and $y = v$ in which $\max(x,y) \leq v$ where v is the minimum of $\max(x,y)$ over the triangle $\mathcal{T}$. There exists a line l that separates the region S from the region $\mathcal{T}$. The line l consists of points (x,y) such that $p_1 x + p_2 y = v$ where p_1 and p_2 are suitably chosen nonnegative numbers satisfying $p_1 + p_2 = 1$. In fact, p_1 and p_2 are the probabilities that define an optimal strategy for Ada. For any point (x,y) representing one of Ben's strategies, the expression $p_1 x + p_1 y$ is the expected payoff. Points on or above the line l satisfy $p_1 x + p_1 y \geq v$. If Ada plays the first row of the payoff matrix with probability p_1 and the second row with probability p_2, then for any point (x,y) in $\mathcal{T}$, that is, for any of Ben's mixed strategies, we have $p_1 x + p_1 y \geq v$. This implies that the expected payoff from each of Ben's strategies is not greater than v, which shows the optimality for Ada of the probabilities p_1 and p_2; and this shows that v is the value of the game.

Games with perfect information

The walking game

It is a beautiful spring day, and Ada and Ben decide on a walk in the country. They have no particular destination in mind, but, to make the trip more interesting, they decide to make a game out of their walk. They decide that Ada will select which way to go at the first fork in the trail, Ben at the second fork, Ada at the third, and, finally, Ben will make a selection at the fourth fork. Depending on where they are at the end of their walk, one of them will give the other a prize in the form of one or more thirst-quenching drinks at Cal's fruit-juice bar. Figure 20.10 is a map of their possible routes showing the prizes (italic numbers) at the end of each route. A positive number is the number of drinks that Ben owes Ada, and a negative number means that Ada treats Ben.

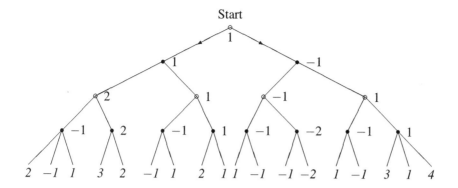

Figure 20.10. The map of the possible routes for the walking game of Ada and Ben. They plan that when they arrive at a node marked with a *hollow* dot, Ada chooses the fork on which to proceed; and at each *solid* dot, Ben makes this selection. The italic numbers in the bottom row represent the payoffs to be made from Ben to Ada at each possible end of the trip. The non-italic numbers do not represent payoffs; they indicate the choices that are made at each node if both players rationally seek their self-interest. These numbers are constructed first at the fourth-level nodes, and then, successively, at the third, second, and first levels. Each number at a *solid* node is the *minimum* of the numbers associated with the possible choices at that node; similarly, each number at a *hollow* node is the *maximum* of choices at that node. These choices are the optimal strategies for Ada and Ben. The number which appears at the *Start* node—it happens to be 1—is the value of the game.

Unfortunately, the process of devising this game takes so much time that the weather starts to change. Alas, such a long walk seems risky in unsettled weather. But then Ada notices that the walk could be shortened without changing the outcome of the game. Ada is willing to concede that Ben's choice at the last fork will be the most favorable one for Ben. For example, if they arrive at the node at the lower left at which Ben has the three choices, 2, −1, and 1, then Ben selects the road that leads to the payoff −1 because that is most favorable to him. Ada's idea is to label the lower left solid node with −1, and to label all possible fourth-level nodes in a similar fashion so that the fourth leg of the journey can be eliminated, thus reducing their exposure to the unsettled weather.

Ben agrees to this plan, but he notes that it is already starting to rain. Ben remarks that one can carry this same process back one more level. At each of the third-level nodes at which Ada makes a choice, Ben concedes that she will make her most favorable choice, and he writes the appropriate number beside each of these four nodes. In fact—it is raining quite hard now—by continuing the process, Ben places a number, as shown in Figure 20.10, at each node—representing the payoff that results if both players play optimally from that node onward. Ben notices that the starting node is associated with the payoff 1. This means that 1 is the value of the game; that Ben can immediately treat Ada to a juice drink of her choice; and that neither of them needs to go out in the rain.

The walking game is an example of a game in *extensive form* as opposed to a *matrix* game. A game is said to be in extensive form if playing the game involves a succession of choices, whereas in a matrix game we suppress the inner structure of the game and consider only the choice of *strategy*.[8] The walking game could be reduced to a matrix game. To do so, we must enumerate all the strategies for Ada and for Ben. In fact, since Ada controls five nodes and each node involves two choices, her total number of strategies is $2^5 = 32$. For Ben there are 10 binary nodes and 2 ternary nodes; therefore, his total number of strategies is $2^{10}3^2 = 9,216$. The payoff matrix for the walking game has 32 rows and 9,216 columns. We will not exhibit this matrix, but if we did it would tell us less about the game than Figure 20.10 does. We showed above that the walking game has optimal strategies and value 1. Furthermore, we showed that the optimal strategies are *pure* (not mixed) strategies. This means that the huge (32 by 9,216) payoff matrix has a *saddle point*.[9]

The diagram in Figure 20.10, ignoring the numbers, is a special case of a mathematical object called a *tree*. The point labeled *Start* is the *root* of that particular tree. (Mathematical trees are frequently drawn, seemingly upside-down, with the root at the top.) The nodes with no successors at the bottom of Figure 20.10 that are labeled with the game payoffs are called the leaves of the tree. A tree consists of nodes and directed line segments connecting them in such a way that starting from the root and traversing the line segments in the defined direction it is possible to reach every node, but it is not possible to traverse a node more than once.

> The folders/directories of a computer disk have a tree structure.

A game like the walking game in which the players move alternately from node to node on a tree with payoffs defined at the leaves of the tree is called a game in extensive form. Games like chess and bridge are games in extensive form. The nodes of the tree are the *states of the game* rather than physical locations. The rules of the game specify how a player can move from one node to another.

The walking game and chess differ from the game of bridge in that the former are games of *perfect information*. A game in extensive form is a game of perfect information if the players know at all times the current state of the game; that is, each player knows what the current node is. Bridge is not a game of perfect information because the players do not know what cards are held by each player. The games of Nim and Kayles previously discussed in Chapter 8 are games of perfect information for which we found optimal strategies.

Since every zero-sum two-person game in extensive form can be reduced, at least in theory, to a matrix game by enumerating the strategies of both players, it is a consequence of the minimax theorem (Theorem 20.6) that every such game has a value and optimal strategies for both players. For games of perfect information, the following stronger assertion is true.

Fact 20.9. *Every zero-sum two-person game with perfect information has a saddle point.*

We verified above that Fact 20.9 is true for the walking game. In fact, in that game we showed that certain pure strategies are optimal and that there is no advantage of playing a mixed strategy. That argument, in which we moved from the leaves of the game tree step by step toward the root, can be extended in a straightforward way to prove Fact 20.9 in general—that is, to show that every game of perfect information has a saddle point.

It would seem to follow that mixed strategies have no benefit in the game of chess. However, chess is so complex that it is unlikely that optimal strategies can be discovered. In practice, chess masters probably mix their strategies for psychological reasons—to surprise and intimidate the opponent.

The game of Hex

Game theory is not able to determine if the first player has an advantage in chess. However, there is a board game, the game of *Hex*, for which it can be shown that a winning strategy exists for the first player.

The proof that in Hex the first player has the advantage was discovered by John Nash. This proof is curious because it is strictly an *existence proof*, and it does not show *how* the first player can win. In practice, the advantage of the first player is not obvious and does not diminish the interest and pleasure in playing the game.

> **Go stones** are lenticular black and white counters (made of slate and shell, respectively) that are used in the traditional Japanese board game of Go.

Hex is played on the diamond-shaped grid of triangles shown in Figure 20.14. The game gets its name from the equivalent form of the game in which the intersections of the lines (the vertexes) in Figure 20.14 are replaced by hexagons, a pattern often seen in ceramic tilework. Figure 20.14 illustrates a 14×14 Hex board. The two players alternate placing, respectively, white and black Go stones on vacant vertexes as shown in Figure 20.11. (Alternatively, the players mark the vertexes with dots of two different colors; a fresh copy of Figure 20.14 is required each time the game is played.) White seeks to form a chain of contiguous white stones connecting the right and left borders of the diamond, and Black seeks to form a chain of black stones connecting the top and bottom of the diamond. Figure 20.12 shows a chain of stones that wins for Black.

Hex was created by the Danish engineer/poet Piet Hein (1905–96) in 1942. It became popular in Denmark under the name Polygon and was played on an 11×11

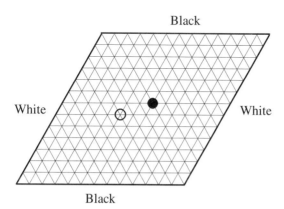

Figure 20.11. The game of Hex. Ada and Ben alternately place, respectively, white and black Go stones on any empty vertex. Ada wins if she forms a connected chain of white stones from the left to the right edges of the board, and Ben wins if he forms a connected chain of black stones from the top to the bottom edges.

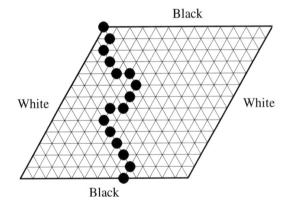

Figure 20.12. A winning chain for Black.

grid of hexagons. The game was rediscovered by John Nash in 1948 when he was a graduate student in mathematics at Princeton. The game was popular in the mathematics common room at Princeton in the 1950s where it was played by many—including the author—on a 14×14 board of hexagons.[10]

The payoffs are 1 if White, the first player, wins; -1 if Black wins; and 0 if there is a draw. However, we will see that in Hex a draw is impossible. Since Hex is a game of perfect information it follows from Fact 20.9 that Hex has optimal pure strategies for both players. We will shortly show that the value of the game is 1; that is, there exists a strategy for White (the first player) that ensures that she wins every time. However, the proof that White has a winning strategy is merely an existence proof; knowing the proof does not help White win the game. In practice, the relative skill of the players is much more important than the order of play.

Proposition 20.10. *In Hex a draw is impossible.*

Proof. If a draw is possible, it means that it is possible to fill the Hex board with white and black stones in such a way that there is neither a winning white chain nor a winning black chain. Suppose that we have such a configuration of stones. Consider totality $\mathcal{T}$ of white stones that are connected by a white chain to the left edge of the Hex board. Since there is no winning white chain, the set of black stones that border $\mathcal{T}$ together with some black stones on the left edge of the board constitutes a winning black chain. Figure 20.13 illustrates this construction.   □

Proposition 20.11. *For Hex there exists a strategy whereby the first player always wins.*

Proof. From Fact 20.9 and Proposition 20.10, either there exists a winning strategy for the first player or for the second player. We will suppose that a winning strategy exists for the second player and derive a contradiction.

If there is a winning strategy for the second player, then the first player makes a random first move and then plays as though he were the second player. In effect, he steals the second player's winning strategy. If the second player's winning strategy

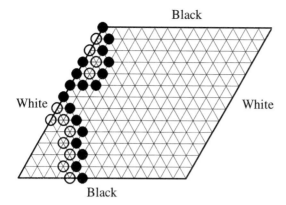

Figure 20.13. If White does not have a winning chain then Black must have one. In this example, the white stones that are connected to the left edge consist of two connected groups. The black stones that border these white stones together with one additional black stone on the left edge of the board constitute a winning chain for Black.

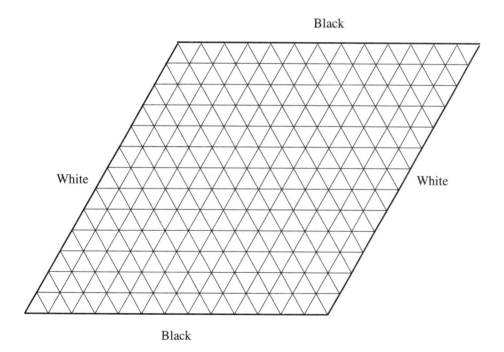

Figure 20.14. Hex board. To play Hex, photocopy this figure; players alternately mark dots in two contrasting colors according to instructions on page 307.

requires him to place a stone where he already placed his initial stone, then he makes another arbitrary move, and so on. These arbitrarily placed stones do him no harm. We assumed that the second player can force a win, and we arrive at the contradiction that the first player can force a win. Therefore, our assumption was wrong. The only other possibility is that the first player can force a win.   □

We have finished the main course, and even the dessert, of our mathematical banquet, but before we go our separate ways, I offer you a mathematical bonbon.

Problem 20.12. Is it possible to use the 31 dominos in Figure 20.15(b) to cover the entire 8×8 chessboard (Figure 20.15(a)) except for the upper left and lower right squares?

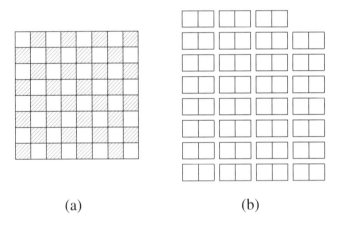

(a)   (b)

Figure 20.15. A chessboard (a) and 31 dominos (b). Each domino is the size of two squares of the chessboard.

Solution. It is not possible to leave two white squares uncovered because each domino must cover a black square and a white square.

So ends our tour! We have experienced many remarkable realizations of the *moment of proof*, the joy of mathematical discovery. For the future, there remains a world of intellectual delights to explore.

Notes

Chapter 1, Reflections

1. Dudeney, 1927, problem 75.

2. To compute the lengths of FS in each of Figures 1.2(c)–(e), we must use the Pythagorean theorem. (See page 172.) In Figure 1.2(e), the distance between S and F is the hypotenuse of a right triangle with legs $32'$ and $24'$. Therefore, we have $\overline{SF} = \sqrt{32^2 + 24^2} = 40'$.

3. See also Polya, 1954, p. 160, problem 13.

4. The horizontal distance between P and P′ is $196''$ and the vertical distance is $147''$. Therefore, by the Pythagorean theorem,

$$\overline{PP'} = \sqrt{196^2 + 147^2} = \sqrt{60,025} = 245''$$

5. Polya, 1954, p. 160, problem 14.

Chapter 2, Hand in Hand

1. For example, in 1887 the German mathematician Richard Dedekind (1831–1916) explored this area in his book, *Was sind und was sollen die Zahlen?*—"What are the numbers and what do they mean?"

2. Hardy, 1967.

Chapter 3, Further Progression

1. See Barnette, 1983, for further details.

2. Lorenz, 1963a, 1963b, and Lorenz, 1964.

3. For further information about chaos, see Gleick, 1987.

Chapter 4, Interesting Numbers

1. Hardy, 1967, p. 37.

2. *Latin:* reduction to an absurdity.

3. That is, the left side is equal to an integral multiple of 4. See Definition 4.5 on page 47.

4. A set $\mathcal{T}$ of numbers (or other mathematical objects for which an order is defined) is said to be *well-ordered* if every nonempty subset of $\mathcal{T}$ has a smallest element. The natural numbers are well-ordered, but the set $\mathcal{D}$ of all nonnegative decimal fractions is not well-ordered because, although zero is the smallest element of the entire set $\mathcal{D}$, the subset of $\mathcal{D}$ consisting of all *positive* decimal fractions (i.e., the subset consisting of all of $\mathcal{D}$ except for the number 0) does not have a smallest element.

5. Sometimes rational solutions are allowed.

6. We have used properties of long division, a very familiar procedure, but we have not given a definition of that algorithm. For an alternate proof, see Ore, 1948, p. 319.

7. The point Q belongs to both the right and left half-lines. This fact is important if Q happens to coincide with the middle point of the three, A, B, and C; in this case, either half-line can serve as h_1.

8. The triangles PQC and BRC are similar triangles in which the sides PQ and BR are corresponding parts. Since triangle PQC is larger than BRC, it follows that d_1 is less than d_0.

9. The statement of Proposition 4.7 does not involve any distances although the proof on page 51 does use distances.

Chapter 5, Tenpins, and Counting

1. See page 230.

2. In the present chapter, we are only interested in *counting* permutations. For the present purpose, it is sufficient to say, for example, that a permutation of the natural numbers 1–15 is one of the $15! = 1,307,674,368,000$ ways of ordering those numbers. However, when we discuss permutations more deeply in Chapter 6, we will need a more subtle definition in which the permutations of 1–15 are *methods* of ordering the natural numbers 1–15.

Chapter 6, Order and Reorder

1. These three problems, first posed by the ancient Greeks, ask for geometric constructions using only a compass and an unmarked straightedge.

Trisecting the angle. Construct an angle one-third the size of a given angle.

Doubling the cube. Construct the side of a cube with twice the volume of a given cube.

Squaring the circle. Construct the side of a square whose area is equal to the area of a given circle.

2. In Euler's time, Königsberg was part of Prussia. Königsberg is now known as Kaliningrad and belongs to the Russian exclave bounded by Poland, Lithuania, and the Baltic Sea. Königsberg was the life-long home of the German philosopher Immanuel Kant (1724–1804).

3. This usage of "image" is a bit different from the common meaning of image as a "faithful representation."

4. Although there are 26 letters in the alphabet, in Definition 6.19 the letters $a, b, \dots, z$ represent an *indefinite* number of elements, not necessarily 26. Furthermore, although Definition 6.19 mentions four elements (a, b, c, and z), a cyclic permutation is permitted to have fewer than four elements. For example, $(1\ 2\ 3)$ is a cyclic permutation of three elements.

5. The element 10 does not occur in any of these three cycles because 10 corresponds to itself—a cycle which optionally could be written (10).

Chapter 7, Outcast

1. See also page 239.

2. Rand Corporation, 1955.

3. For more details, see Computational Science Education Project, 1995. See Press, Flannery, Teukolsky, and Vetterling, 1988, for a list of other "good" LCGs. The following LCG is recommended in Park and Miller, 1988:

$$x_{n+1} = 16807 \cdot x_n \pmod{p}$$

where p is equal to $2^{31} - 1 = 2,147,483,647$.

Chapter 8, The Power of Two

1. Ball and Coxeter, 1987, p. 36.

2. Note that, since duplicate piles are permitted, $\{6,6\}$ is a meaningful Nim position. This illustrates that our usage of curly brackets for Nim positions differs from the standard usage of curly brackets for defining sets by enumeration of their elements. In the latter meaning, duplicates are not allowed. See page 138.

3. Dudeney, 1927, problem 73, describes Kayles with some historical remarks concerning the bowling game of Kayles. The preliminary observations concerning the theory of Kayles in Ball, 1926, are updated in later editions of that work, e.g., Ball and Coxeter, 1987. The theory that underlies Figure 8.9 was first published in Sprague, 1935–36. The information in Figure 8.9 was first published in Guy and Smith, 1956. The games of Nim and Kayles together with many other examples and far-reaching generalizations are discussed in Berlekamp, Conway, and Guy, 1982.

4. The Rosetta stone, discovered in 1799 in the Egyptian town of Rosetta, enabled the decipherment of Egyptian hieroglyphics in 1822 by the 32-year-old French Egyptologist Jean François Champollion (1790–1832).

5. Theorem 3.13.

6. Gleick, 1987.

7. Devaney, 1995, describes the Chaos Game and recommends it as a device for introducing fractals in both middle school and high school classrooms.

8. Devaney, 1995, recommends that a group of students play Chaos Games starting with the same triangle on separate acetate sheets. When the sheets are layered and projected, the Sierpinski triangle appears. This procedure has the advantage that a large number of points can be marked with little individual effort.

9. For more information on Nim, Kayles, and many other similar games see Berlekamp et al., 1982. For more about fractals, see Barnsley, 1993.

Chapter 9, Divide and Conquer

1. In computer applications, the number of data moves for these sorting methods depends on the technical manner in which the list of items is implemented. The data structure known as *linked list* leads to an analysis similar to that of playing cards.

2. For example, FORTRAN 90 supports recursion, but FORTRAN 77 does not.

3. Hoare, 1962.

4. We have just scratched the surface of a very large subject. The reader interested in reading more about computer algorithms could consult Aho, Hopcroft, and Ullman, 1983.

5. The FFT algorithm was discovered by Cooley and Tukey; see Cooley and Tukey, 1965.

Chapter 10, Set and Match

1. Mapping = function. See page 46.

2. See page 116. This defines the sense in which Cantor's middle-thirds set is very large. We have previously discussed aspects in which, paradoxically, Cantor's set is also very small.

Chapter 11, Chance Encounter

1. Ore, 1953.

2. Todhunter, 1949.

3. Selvin, 1975.

4. By *birthday* we mean, more precisely, *birthday anniversary.*

5. Or 367 if we are concerned that some member of the group is born on a leap year on February 29.

6. Dubins and Savage, 1976.

7. The probability of winning a game of frustration solitaire is $0.01623\ldots$, but the details of the solution are beyond the scope of this book. See Doyle, Grinstead, and Snell, 1994.

8. Todhunter, 1949.

9. There is more than one answer to this question. See, for example, the essay by John Maynard Keynes, in Newman, 1956, p. 1368.

10. The interested reader could learn more about this engrossing and useful subject in Parzen, 1992.

Chapter 12, Cutouts

1. Dudeney, 1927, problem 26.

2. Dudeney, 1958, pp. 28–35.

3. Ball and Coxeter, 1987, p. 92.

4. Theorem 16.4 on page 228.

5. Loomis, 1940.

6. In number theory, *congruent* has a different meaning. See Definition 7.3 on page 94.

7. For a proof of this fact, see Ore, 1948.

8. For further information on double-angle triangles, see Brown, 1948. According to Dickson, 1952, pp. 213–4, double-angle triangles were first discussed by K. Schwering in 1886.

Chapter 14, Two Pearls

1. Cicero's discovery of Archimedes' tomb has inspired paintings of the event by Pierre Henri de Valenciennes (1787) and Benjamin West (1797, 1804); and an engraved frontispiece by Xaver Weinzierl (1806) in a German translation of *Tusculan Disputations*. There are two sites in Syracuse, Sicily, claiming to be Archimedes' tomb, but one of them, dating from a time much later than Archimedes, is clearly false. The column with the sphere and cylinder described by Cicero has apparently disappeared.

Chapter 15, New Numbers for Old

1. In the sense defined on page 93.

2. Equivalence classes are generated by an equivalence relation that satisfies reflexivity, symmetry, and transitivity. See the discussion on page 139.

3. Scipione del Ferro (1465–1526), Niccolò Tartaglia (the Stammerer) (c. 1499–1557), Girolamo Cardano (1501–76), Lodovico Ferrari (1522–65), and Rafael Bombelli (c. 1526–73). See *An Extraordinary and Bizarre Story* in Eves, 1980, concerning intrigue and deceit over the methods of solving cubic and quartic equations.

4. For example, the quaternions of William Rowan Hamilton and the octonians of Arthur Cayley (1821–97).

5. See Mandelbrot, 1982. For a collection of stunning images based on or related to the Mandelbrot set, see Peitgen and Richter, 1986.

Chapter 16, Prime News

1. For a proof of the unique factorization theorem, see Ore, 1948, p. 51. The theorem is not as obvious as it may seem because there are systems that resemble the natural numbers in most respects but lack unique factorization.

2. See page 61.

3. Lectures presented at Stanford University during the fall quarter of 1950 by J. G. van der Corput.

4. Also see page 182 for a discussion of Fermat's last theorem.

5. The 10 in 10-*perfect* refers to the sum of *all* divisors including the number itself.

Chapter 17, The Unknown Division

1. Quoted in Ore, 1948.

2. See Selby, 1986, for further information.

3. See Beeler, Gosper, and Schroeppel, 1972, Item 101B.

4. The variable N represents a name, e.g., "Smith."

5. See also page 223.

6. For further reading on these topics, see Ore, 1948. For a treatment of continued fractions beyond the scope of this book, see Hardy and Wright, 1979.

Chapter 18, Secret Messages

1. RSA is named after Rivest, Shamir, and Adleman, who devised this cryptosystem.

2. Example 17.15 can be used both here and in Example 18.11 because $\phi(77) = \phi(61) = 60$.

3. If Ada, Ben, and Cal share the same modulus, they might be able to crack each other's deciphering keys because each might know the two prime factors of the modulus. Moreover, each knows the other's public enciphering key. We have seen that one can find the deciphering key corresponding to *any* enciphering key relative to a modulus with known prime factors.

4. For further information concerning applications of number theory to cryptography, see Koblitz, 1994.

Chapter 19, Be Wise, Optimize

1. Dantzig, 1963.

2. Stigler, 1945.

3. Dantzig, 1963.

4. Hitchcock, 1941.

5. For the technical details of the simplex method, see Dantzig, 1963.

6. See page 129.

7. Karmarkar, 1984.

8. Kuhn, 1955.

9. Kőnig, 1950.

10. For more details on linear programming, see Dantzig, 1963.

Chapter 20, The Play's the Thing

1. Confucius, 1997. (Translated by Charles Muller.)

2. Von Neumann and Morgenstern, 1944.

3. See page 149.

4. See page 160.

5. See Williams, 1986.

6. Von Neumann, 1928.

7. See page 186 and Figure 13.1.

8. In the sense defined on page 293.

9. See Definition 20.4, page 298.

10. See Gardner, 1988, for more details of the history of Hex.

References

Aho, A. V., Hopcroft, J. E., and Ullman, J. D. 1983. *Data structures and algorithms.* Reading, MA: Addison-Wesley.

Ball, W. W. R. 1926. *Mathematical recreations and essays* (10th ed.). London: Macmillan.

Ball, W. W. R., and Coxeter, H. S. M. 1987. *Mathematical recreations and essays* (13th ed.). New York: Dover Publications. (First published in 1894.)

Barnette, D. 1983. *Map coloring, polyhedra, and the four-color problem.* Washington, DC: Mathematical Association of America.

Barnsley, M. F. 1993. *Fractals everywhere* (2nd ed.). Boston: Academic Press.

Beeler, M., Gosper, R. W., and Schroeppel, R. 1972. *HAKMEM* (Tech. Rep. No. 239). Massachusetts Institute of Technology A. I. Laboratory. (Available on the World Wide Web.)

Berlekamp, E. R., Conway, J. H., and Guy, R. K. 1982. *Winning ways for your mathematical plays* (vols. 1 and 2). London: Academic Press.

Brown, D. M. 1948. Numerical double-angle triangles. *The Pentagon, 8,* 74–80.

Computational Science Education Project. 1995. Random numbers. *World Wide Web.* (http://npac.syr.edu/projects/csep/m/m.html)

Confucius. 1997. Analects. *World Wide Web.* (Translated by Charles Muller, http://www2.gol.com/users/acmuller/contao/analects.htm)

Cooley, J. W., and Tukey, J. W. 1965. An algorithm for the machine computation of complex Fourier series. *Math. Comp., 19,* 297–301.

Dantzig, G. B. 1963. *Linear programming and extensions.* Princeton, NJ: Princeton University Press.

Devaney, R. L. 1995. Chaos in the classroom. *World Wide Web.* (http://math.bu.edu/DYSYS/chaos-game.html)

Dickson, L. E. 1952. *History of the theory of numbers* (vol. 2). New York: Chelsea.

Doyle, P., Grinstead, C., and Snell, J. L. 1994. Frustration solitaire. *World Wide Web*. (http://math.ucsd.edu/~doyle)

Dubins, L. E., and Savage, L. J. 1976. *Inequalities for stochastic procesess: How to gamble if you must*. New York: Dover.

Dudeney, H. E. 1927. *The Canterbury puzzles* (2nd ed.). London: Thomas Nelson and Sons.

Dudeney, H. E. 1958. *Amusements in mathematics*. New York: Dover. (Originally published in 1917.)

Eves, H. 1980. *Great moments in mathematics (before 1650)*. Washington, DC: Mathematical Association of America.

Gardner, M. 1988. *Hexaflexagons and other mathematical diversions*. Chicago: University of Chicago Press. (The First Scientific American Book of Puzzles and Games. First published in 1959.)

Gleick, J. 1987. *Chaos: Making a new science*. New York: Penguin Books.

Guy, R. K., and Smith, C. A. B. 1956. G-values of various games. *Proceedings of the Cambridge Philosophical Society, 52*, 514–526.

Hardy, G. H. 1967. *A mathematician's apology*. London: Cambridge University Press. (With a foreword by C. P. Snow.)

Hardy, G. H., and Wright, E. M. 1979. *An introduction to the theory of numbers* (5th ed.). Oxford: Oxford University Press.

Hitchcock, F. L. 1941. The distribution of a product from several sources to numerous localities. *J. Math. Phys., 20*, 224–230.

Hoare, C. 1962. Quicksort. *Computer Journal, 5*(1), 10–15.

Karmarkar, N. K. 1984. A new polynomial-time algorithm for linear programming. *Combinatorica, 4*, 373–395.

Koblitz, N. 1994. *A course in number theory and cryptography*. New York: Springer-Verlag.

Kőnig, D. 1950. *Theorie der endlichen und unendlichen Graphen*. New York: Chelsea. (Reprint. Originally published in 1936.)

Kuhn, H. W. 1955. The Hungarian method for the assignment problem. *Naval Research Logistics Quarterly, 2*, 83–97.

Loomis, E. S. 1940. *The Pythagorean proposition* (2nd ed.). Ann Arbor, MI: Privately printed, Edwards Brothers.

Lorenz, E. N. 1963a. Deterministic nonperiodic flow. *Journal of Atmospheric Science, 20*, 130–41.

Lorenz, E. N. 1963b. The mechanics of vacillation. *Journal of Atmospheric Science*, *20*, 448–64.

Lorenz, E. N. 1964. The problem of deducing the climate from the governing equations. *Tellus*, *16*, 1–11.

Mandelbrot, B. 1982. *The fractal geometry of nature*. San Francisco: W. H. Freeman.

Neumann, J. v. 1928. Zur Theorie der Gesellschaftsspiele. *Mathematische Annalen*, *100*, 295–320.

Neumann, J. v., and Morgenstern, O. 1944. *Theory of games and economic behavior.* Princeton, NJ: Princeton University Press.

Newman, J. R. (ed.). 1956. *The world of mathematics*. New York: Simon and Schuster.

Ore, O. 1948. *Number theory and its history*. New York: McGraw-Hill.

Ore, O. 1953. *Cardano, the gambling scholar.* Princeton, NJ: Princeton University Press. (Contains a translation from the Latin of Cardano's *Book of Games of Chance*, by Sydney Henry Gould.)

Park, S., and Miller, K. 1988. Random number generators: Good ones are hard to find. *Transactions of the ACM.*

Parzen, E. 1992. *Modern probability theory and its applications.* New York: Wiley.

Peitgen, H.-O., and Richter, P. H. 1986. *The beauty of fractals: Images of complex dynamical systems.* Berlin: Springer-Verlag.

Polya, G. 1954. *Mathematics and plausible reasoning,* vol. 1, *Induction and analogy in mathematics.* Princeton, NJ: Princeton University Press.

Press, W. H., Flannery, B. P., Teukolsky, S. A., and Vetterling, W. T. 1988. *Numerical recipes in C.* Cambridge: Cambridge University Press.

Rand Corporation. 1955. *One million random digits with 100,000 normal deviates.* Glencoe, IL: The Free Press.

Rivest, R. L., Shamir, A., and Adleman, L. M. 1978. A method for obtaining digital signatures and public key cryptosystems. *Communications of the ACM, 21*(2), 120–126.

Selby, P. 1986. *Analytic geometry.* New York: Harcourt Brace Jovanovich.

Selvin, S. 1975. A problem in probability. *American Statistician, 29*(1), 67.

Sprague, R. P. 1935–36. Über mathematische Kampfspiele. *Tôhoku Journal of Mathematics, 41*, 438–44.

Stigler, G. J. 1945. The cost of subsistence. *J. Farm Econ., 27*(2), 303–314.

Todhunter, I. 1949. *A history of the mathematical theory of probability.* New York: Chelsea. (Reprint. Originally published in 1865.)

Williams, J. D. 1986. *The compleat strategyst.* New York: Dover.

Index